P9-CAB-114

GENOMIC MESSAGES

GENOMIC MESSAGES

HOW *the* EVOLVING SCIENCE *of* GENETICS AFFECTS OUR HEALTH, FAMILIES, *and* FUTURE

GEORGE ANNAS, J.D., M.P.H.
SHERMAN ELIAS, M.D.

HarperOne
An Imprint of HarperCollins*Publishers*

GENOMIC MESSAGES: *How the Evolving Science of Genetics Affects Our Health, Families, and Future.* For information address HarperCollins Publishers, 195 Broadway, New York, NY 10007.

HarperCollins books may be purchased for educational, business, or sales promotional use. For information please e-mail the Special Markets Department at SPsales@harpercollins.com.

HarperCollins website: http://www.harpercollins.com

FIRST EDITION

Library of Congress Cataloging-in-Publication Data is available upon request.

ISBN 978-0-06-222825-3

15 16 17 18 19 RRD(H) 10 9 8 7 6 5 4 3 2 1

To our grandchildren

Contents

Introduction

Genomics has captured the attention of presidents and physicians, of science enthusiasts and health-conscious Americans. *Genomic Messages* is about what your genome, as read and interpreted by a skilled geneticist, can tell you about your health and your family's health, today and in the near future. Genomics will change how we think about ourselves and our fellow humans, and is powerful enough that it could transform American medicine in the coming decades. Because of its transformational potential, it is critical that the evolving science of genomics is introduced into medical care in a way that makes the health system better and more responsive to patients, improves communication and the physician-patient relationship, improves overall health, and contains cost, all the while avoiding the past pitfalls of the old genetics: eugenics, stigmatization, and discrimination. That's a big order!

A genomics that improves our medical care and our lives is only possible if we as citizens, consumers, and patients all critically engage with the science of genomics. Critical engagement is possible because genomics can and should be made accessible to non-specialists and the "lay public." It is also a good time for public engagement because genomics has so far been actively introduced into clinical medicine primarily in two areas: prenatal screening and cancer research. Cancer research is also at the core of the genomic medicine initiative which President Obama announced in his 2015 State of the Union address, relabeling personalized medicine, "precision medicine." As the president later put it, jump-starting genomic research with new federal funding is one of his few budget proposals that has strong bipartisan support. A major component of the president's new genomic initiative is the planned construction of a massive DNA data bank with a million Americans sharing their personal, medical, and genomic information. This data bank can only be built and used, the president underscored, if the privacy of the participants can be credibly protected. Four of the ten chapters of this book specifically address each of these areas: precision (genomic) medicine, prenatal screening, cancer treatment, and genomic privacy. The other six are interconnected: the nature of genomic information, nature and nurture, pharmacogenomics, reprogenomics, children, and "genomics future." One of the themes of this book is that the coming flood of genomic information is likely to make at least some of your medical treatment more "precise" and "personalized," but this flood of genomic information will also bring you and your physician new levels of uncertainty. The cliché is true: we do not know the future, and even with genomics we can no more predict our future health with certainty than we can predict the weather, or even the next terrorist attack.

We are a health lawyer-bioethicist (George) and an obstetrician-gynecologist-geneticist (Sherman) team who have worked together on the clinical aspects of genomics for more than three decades, during which we have published academic books and articles, participated in national and international clinical and legal programs, and worked together on ethics panels. Our working partnership is unusual. Physicians routinely see lawyers as predators, and themselves as their prey. On the contrary, our work together has reinforced our view that cooperation between medicine and law, and genomics and society, is an essential ingredient for progress. Only a genomic medicine that accounts for the values and concerns of the public and patients, is likely to produce useful, accessible, and affordable innovations that can improve your life and health. *Genomic Messages* contains many stories of real patients. These stories are from Sherman's own patients (with identities masked), cases published in the medical literature, and celebrity patients who have made their own stories public.

When we began *Genomic Messages*, Sherman had no reason to even suspect that he might not live to see it published. But shortly after the manuscript was completed, Sherman died. His ideas, of course, live on in this book, and in the lives of his family, colleagues, and the patients he cared for. George dedicates *Genomic Messages* to Sherman; but Sherman and George had agreed to dedicate this book to the future: to our grandchildren.

GENOMIC MESSAGES

CHAPTER 1

The Coming Flood of Genomic Messages

The replication of DNA is a copying of information. The manufacture of proteins is a transfer of information: the sending of a message.

—James Gleick, *The Information* (2011)

We are surrounded by genomic messages much the way fish are surrounded by water. Like fish, we pay little or no conscious attention to these messages. Our bodies, nonetheless, are constantly interpreting the messages, and the way our genes interact with each other and our environment determines the state of our health. These messages are contained in the most remarkable molecule in nature—deoxyribonucleic acid, or DNA. DNA contains the instructions for human development, survival, and physiologic functions, as well as ensuring that our biological information will be passed to our children and future generations.

Some genomic messages are visible on the surface of our bodies, including facial structure, skin color, and height. Other messages predispose us to illnesses that might be translated into diseases later in our lives. Genomic messages can also be read from DNA samples taken from fetuses and newborn babies. Whether translated, mistranslated, or ignored, our DNA and the messages we derive from it will affect (together with our environment) how we live and how we die, as well as the health and future of our siblings, children, and grandchildren. Some of us will live long, others of us will die young; some of us will develop colon or breast cancer, others will not; some of us will suffer from premature dementia, others will age with minds fully functional. Of course, the probability of death is still 100 percent for us humans. But we may be able to lengthen our lives and improve their quality, and those are worthy goals for medicine. The good news is that we may be able to prevent or treat some diseases by reading the genomic messages in our DNA. For the immediate future, however, we will only be able to probabilistically predict, but not prevent, most diseases.

This book will help you make your own decisions about whether and how to use the evolving genomics in your own life. To the extent that health insurance companies, the government, and even your employers think that genomics can save them money, they are likely to pressure you to use the new genomics. Also, to the extent that private corporations believe they can make money by getting you to use the new genomics, you will be subjected to commercials every bit as pervasive as current prescription drug advertising on TV. In this book we will tell you what we think and why, but we will strive to be as objective as we can to help you decide whether to embrace or reject invitations to genomic interventions. Our goal is to enable you to be a more informed and critical consumer of the evolving world of genomics that will invariably affect you, your family, and your ethnic community, perhaps profoundly.

Your ability to benefit from genomics will also depend, at least in part, on decisions made by physicians, the medical profession, hospitals, biotech and pharmaceutical companies, and public health officials—so we will suggest how they can act to most effectively and reasonably make the fruits of genomics available to the public, and how legislation and regulation could maximize the benefits of genomics and minimize its dangers. Every literate citizen in the United States will soon need to be familiar with genomic medicine, research, and privacy, and importantly, our rights and the limitations of government and corporate access to our DNA and that of our children.

Genomic Information

Most people use *genetics* and *genomics* interchangeably. This is consistent with both popular and scientific usage. Nonetheless, technically, *genetics* refers to the action of single genes, whereas *genomics* refers to the totality of our DNA and its interaction with the environment, and the broader term will ultimately replace the narrower one for most purposes. In most contexts, we will use *genomics,* but in some contexts, especially historical, *genetics* will be more accurate. Genomic messages are already beginning to change the practice of medicine and have the power to radically alter how your physician thinks about you and thus how your physician will talk to you and treat you and your family. Genomic messages will also likely transform what we now think about medical privacy, and could even affect the legal and ethical doctrine of informed consent.

How much of our medical future can our DNA tell us, and how much do we really want to know? How will our increasing knowledge of genetics and genomics change what medicine can do for us or how we think about our lives, our families, and ourselves? How can each of us take advantage of the coming flood of genomic information without getting drowned in information? Will, for example, genomic information simply overwhelm modern medicine by its sheer "big data" complexity? As medical historian Hallam Stevens put it, biology is "already obsessed with data. . . . [Our goal should] not be to be swept up in this data flood, but to understand how data mediate between the real and the virtual." Stevens explains that digitalized genomic messages can change the way we think about life itself as "data bring the material [DNA] and the virtual [digital] into new relationships." This is because, as he puts it, "data properly belong to computers—and within computers they obey different rules . . . and can enter into different kinds of relationships." One way to think about the different kinds of relationships is to contrast the chemical language of DNA's double helix with its digital representation in computer language. Exactly where digitalized and decoded DNA informatics takes us will depend to a large extent on our ability to interpret it and the ways we decide to use it. It is not inevitable that digitalized DNA will let us construct a "stairway to heaven," or even a better life here on Earth.

Having digitalized DNA, the goal now is to convert the resulting electronic information (data) into knowledge of biology, both population biology and individual biology. Most of us are much less interested in the risk an average American has of obesity than we are in our personal risk, and the risk faced by our children. But for the vast majority of diseases, population average risk is all we have to go on. Nonetheless, many of us will likely want to know whatever our genomes and the genomes of our children can tell us about our probabilistic medical future, at least if there is some action we can take to improve it. On the other hand, because genetic information can alter the way we think about ourselves and our future in both positive and negative ways, some of us will not want to participate in this new world of genomics, just as many of us choose not to participate in annual physical exams or various forms of cancer screenings. There is a risk that we could come to think of ourselves as sick—even though we are completely healthy—because we are at heightened genomic risk to get a disease in the future. Seeing ourselves and our children as born diseased and destined to suffer—rather than as born healthy and destined to live a healthy life, would, we think, be a major human tragedy. That is just one reason why as the quantity of genetic information grows, the right not to know will become as important as the right to know. The right not to hear the genetic messages that could be conveyed by our DNA is not a "right to be ignorant" but a right to live our lives as we see fit. It is a right that is fundamental to informed consent, and it applies to genomics just as it applies to all other areas of medical practice.

Genomics is technologically driven, and we will provide introductions to the major technologies that are driving genomics, including computer technology, IVF, noninvasive prenatal screening, cloning, and genome editing. All technologies change the way we think by changing what we can do. Genomic technologies are so powerful that they have changed the way we think about ourselves and our future even before they have substantially changed what we and our physicians can do. The world of medicine is just beginning to incorporate genomic information into medical practice, and the evolving use of genomic information will ultimately change the practice of medicine itself, at least once your whole genome sequence is made part of your electronic health record. Changes in medical practice will include the tests your physician will want to perform on you, the drugs that can be safely prescribed for you, and the actions, such as diet and exercise, that your physician may suggest you take to reduce your risk of specific diseases.

For now, one of the major challenges posed by genomic information is its sheer size, only hinted at by the phrase *big data*. Our DNA has been described as a master blueprint, a musical score, and even a data bank. But perhaps the most useful and widespread analogy is to think of your DNA as a recipe. The way your cake comes out depends not just on the recipe but on the ingredients—their quality, quantity, and how they are mixed and prepared. Nonetheless, the most common metaphor remains the book. You could think of DNA as the "book of life," or even as your biography. A DNA data bank (a collection of genomes from hundreds or even millions of people, stored on one or more servers) can be thought of as a library, like the imaginary library of the Argentine fable writer Jorge Luis Borges. Borges describes an infinite library that contains not only every book ever written but every book that could possibly be written—in every combination of letters and words. The fictional library is both completely inclusive and completely incomprehensible.

Our genomes are currently much like the books in Borges's library, each holding an incredible amount of complex information contained within 3 billion tiny bits of paired code, called DNA, orderly arranged within a tightly wound double-helix formation. Like the letters of a book, they contain seemingly infinite combinations composed of four chemical bases (which can be thought of as composing a four-letter alphabet): adenine, thymine, cytosine, and guanine (abbreviated *A, T, C,* and *G,* respectively). The sequences of these letters are responsible for the formation and development of almost all living organisms, as well as for preserving genetic information from generation to generation, and for cell function. Even this description is inadequate: the DNA molecule is not linear, but is bunched together in loops and folds. This means that a particular strand of DNA may be in physical contact with another strand that is millions of letters away, and this contact may affect its function. The U.S. Supreme Court's definition of DNA is so scientifically accurate we have included it as Appendix A. Having adopted the language metaphor, as James Gleick has noted, it seemed natural for biologists to also adopt related concepts, including "*alphabet, library, editing, proofreading, transcription, translation, nonsense, synonym,* and *redundancy.*"

Unlike the books in the Borges library, which can never be given meaning, we are slowly learning how to read our DNA and translate or "decode" the genetic messages contained in the approximately 22,000 genes in our forty-six chromosomes. This is being accomplished primarily by collecting and comparing vast numbers of individual genomes. Interpreting what genetic messages mean for you is currently the most challenging aspect of genomics. This is because genes interact in ways we do not understand, and our internal and external environments directly affect how our genes express themselves. Gene expression, for example, can be controlled by "switches" in the noncoding regions of the genome which can turn genes on or off. Another major influence on our genes is our "microbiome." We are home to 100 trillion microbes (bacteria, yeasts, parasites, and viruses), which affect whether and how our genes are expressed. Until recently it was also assumed that our DNA was stable and that its functioning could not be easily modified. We now know that environmental factors modify the functioning of genes, and this has enabled a new scientific area of research, epigenetics ("on top of" or "over" genetics).

Medicine is still very early in the genomic quest for a longer life, as well as the quests to cure or prevent Alzheimer disease, Parkinson disease, diabetes, or cancer by attacking their genetic roots. The massive project to develop an Encyclopedia of DNA Elements, known by the acronym ENCODE, for example, in 2012 published its first results, which described functional elements (other than our 22,000 protein-coding genes) that make up the human genome. It appears there is very little "junk," or nonfunctioning DNA. "The ENCODE consortium has assigned some sort of function to roughly 80% of the genome, including more than 70,000 promoter regions . . . and nearly 4,000,000 enhancer regions that regulate expression of distant genes."

Genomics leader Eric Lander of MIT has described the current state of genomics using another metaphor, a map: "It's Google Maps. . . . [T]he human genome project was like getting a picture of Earth from space. It doesn't tell you where the roads are, it doesn't tell you what traffic is like at what time of the day, it doesn't tell you where the good restaurants are, or the hospitals or the cities or the rivers. . . . My head explodes at the amount of data." We're with Lander in marveling over the vastness of information that is being added to genetic messages, as well as the effort that will have to be devoted to deciphering and interpreting them.

In February 2015, after it was determined that gene "switching areas" in the genome could turn genes on and off, Lander commented that it was extremely complicated to figure out which switches went with which genes. Boston was still digging out from a series of major snow storms that crippled the city's transportation system, and Lander used the Boston subway system as his new metaphor. He thought it would be possible to figure out which subways lines were disrupted by the storm by determining which employees were late for work. Similarly, when a genetic circuit is shut down, Lander thought it possible to determine which genes were affected, and thus which genes are likely to be associated with the circuit. The name of the new project is the Roadmap Epigenomics Project, which the researchers involved described as an effort to construct a "road map to the human epigenome (a collection of chemical modifications of DNA that alter the way genetic information is used in different cells). This is powerful new research, but as the editors of *Nature* put it in announcing some of the results, "despite the progress, each question that the genome helps answer throws up further questions. Much remains to be understood about how genetic information is interpreted by the individual cells in our body." All of this confirms our initial intuition: scientists are early in the genomics research phase, and many if not most clinical applications remain in the distant future. For the immediate future, we are confronted with one of what former secretary of defense Donald Rumsfeld described as the "known unknowns," things we know we don't know.

Your genes are a vital part of you, but you are much more than just your genes, more than even your entire genome. This means we will never be able to understand human life or humanity no matter how much we understand about our genome; humans simply do not live their lives on the genetic or molecular level. Nonetheless, the more we discover about our genomes, the more difficult it becomes to resist thinking that the more we know about the tiny parts that make up our DNA, the more we will know about ourselves and our lives. This is evident whenever someone defines a person or a fetus based only on a specific genetic characteristic. We have already lowered the cost of whole-genome sequencing for research to $1,000, and this (or less) will likely be available in the clinic soon. The $1,000 genome has always seemed like a reasonable technological goal, and a necessary one to bring the genome into clinical medicine by pricing it on the level of an MRI.

At the clinical level, however, the decreasing price of a genomic sequence has so far primarily produced more complex translation questions. Our current situation is sometimes described by the only half-joking observation that we will soon have "the $1,000 genome with the $1,000,000 interpretation." This is a purposeful exaggeration but it underlines two points. The first is that cost alone cannot determine use. The famous story of the $5 elephant makes the point: you would not buy an elephant, even for only $5, if you did not want an elephant. The elephant is much more trouble to most people than its price alone would suggest. The second point is that regardless of price, interpreting genetic information is much more difficult than collecting it. This is the primary reason why companies in the United States, China, and Europe are collecting genomes from tens of thousands of people: to do research on these collections to identify genetic sequences that matter to health. It is also why President Obama called for a new project to collect DNA and medical records from a million Americans in 2015. Collecting genomic data is, of course, a means to an end (better health), not a goal in itself. Stockpiles of genomic information alone will not help anyone and could hurt us all by enabling genetic discrimination. We will need to shift the focus of our research projects from simply collecting and sequencing DNA to figuring out how, like the "switching" research, genomic information can be used to help us.

In her futuristic *MaddAddam* trilogy, Canadian novelist Margaret Atwood imagines a different kind of flood of information, a "waterless flood," in which a lethal pandemic of a bioengineered virus destroys most life on the planet. Atwood's cautionary tale reminds us that genetic information has a dark side. We have properly begun to take steps to regulate plague-related research, such as research designed to make a virus more virulent or deadly, for our own protection and that of the planet. We will address the regulation of international research in chapter 10, but mostly this book is about helping you make your own decisions about using the evolving science of genomics in your own life.

To take full advantage of the evolving genomics, you will need to know more than just the scientific and medical aspects—you'll also need to know the relevant legal and ethical aspects. Physicians and lawyers must work together in this realm. Although often seen as natural enemies, even as prey and predator, we believe that not having doctors and lawyers working together is counterproductive and shortsighted. Just as genetics cannot be isolated from medical practice, so too medical practice and genetics cannot be understood without an appreciation for the legal and ethical issues they raise. This is true not only in the courtroom and legislative hearing room but at the bedside as well. The intimate relationship of medicine and law in genomics is perhaps most apparent in the realm of what has come to be known as genomic privacy.

Genomic Privacy

Both physicians and lawyers have historically protected privacy. We believe that your genome, which George has called your "future diary," should be considered as private as your diary. No one should be able to "open" it or "read" it without your authorization. To put it another way, your genome is so personal and important to how you view yourself, and potentially to how others view you, that you should always be considered the owner and person in charge of your genome and the information it contains.

The idea for the "future diary" metaphor came from the late *New York Times* commentator William Safire, who argued that diaries should remain private because they are uniquely our own. We keep a diary "to reveal our youthful selves to our aging selves." We think Safire is correct and that his reasoning applies to our genomes as well: we open our genomes "to inform our younger selves about our aging selves," and only we should be able to determine if our "future diary" will be opened and read. Genomes can also be used by individuals to help them identify risk factors—but this suggests a less benevolent metaphor: the DNA profile as a "personalized health threat matrix" that identifies the conditions most likely to kill or sicken us.

Neither of these metaphors means that we think your DNA alone is capable of telling a coherent story about your life, or even your health. We agree with linguist Ann Jurecic that "genome sequences aren't like stories. . . . [T]here is a profound difference between genetic data and a story that seeks to define a life's meaning." She believes making sense of our interactions with genomic information "will require experimentation with new literary forms" that will enable us to tell stories about ourselves not focused on the molecular level but entangled with the whole "earth in which we live." Another writer, Christine Kenneally, has eloquently argued that our DNA tells us more about our past than our future. You are a product of humanity's history. "The millions of bits that initially made you—all the cultural bits and the genetic bits, each with its risk factors, predispositions, and probabilities—were shaped by the past." She seems right about this, and the "past and future diary" may be a better metaphor for DNA privacy (figure 1.1).

Our DNA's inability to define us, however, should not make it a public resource, any more than it makes our blood or organs public resources. Instead, society should take genomic privacy seriously enough to outlaw the collection of an individual's DNA for testing without authorization. This should be done not only because knowledge of one's genome can probabilistically predict at least some of our future health problems, but also because DNA information can be easily distorted and used against us. This potential for misuse and stigmatization is also why we remain surprised that no one has tried to use the DNA of a presidential candidate (taken from a drinking glass) against the

1.1 Artist's rendition of George's DNA "Future Diary" in USA *Today*. Marcia Staimer, "Future diary," in M. Snider, "Genetic Privacy Laws Sought," *USA Today*, November 17, 1993, 9D.

candidate. An opponent could suggest, for example, that the candidate is genetically predisposed to a condition that could adversely affect his judgment, such as dementia. This tactic can be labeled "genetic McCarthyism." This misguided use of genetics is in the same category as trying to identify a "mass-murderer gene." For example, University of Connecticut researchers wanted to examine the DNA of Adam Lanza—the killer of twenty-eight people in Newtown, Connecticut—to see if they could identify a "mass-shooter gene." We are easily seduced by genomics and need to keep our common sense fully engaged when genomic answers to complex human behavior questions are suggested. DNA-centric thinking is an example of what psychologist Daniel Kahneman termed "thinking fast," when what is called for in genomics is "thinking slow."

As important as genomic privacy is to political candidates, it is likely that you will find it much more important to yourself, your family, and your physician. Genomic privacy is really about control of your genomic information, including whether to seek it at all and with whom to share it. Sharing your DNA, even with your family, should, we will argue, be a personal choice. But we also think it should be an educated and well-considered choice. Current federal law prohibits discrimination on the basis of genetics in employment and health insurance but not in life, long-term care, or disability insurance. Of course, the importance of genomic privacy is based on medicine's ability to extract and interpret messages contained in your genome, and that ability is improving daily.

Genomic Medicine

The ability to sequence an individual's entire genome (whole-genome sequencing, or WGS) has also opened a door to a whole new field of medicine known as "personalized medicine," "precision medicine," or simply "genomic medicine." By sequencing an individual's genome, physicians will be able to obtain a genetic profile to guide them in diagnosis, treatment, and possibly even prevention of disease. Some genomic messages are already informing medical treatment, particularly in certain cancers. As an overview, we briefly discuss four examples, one involving a medical treatment based on a genetic finding, another involving a series of preventive measures taken based on a genetic finding, a third in which the primary result of a genetic finding was uncertainty, and a final example of noninformative genetics: Angelina Jolie Pitt; Sergey Brin; Donna, one of Sherman's patients; and Robert Green, a colleague of George and Sherman.

Genetic screening for cancer-predisposing genes, coupled with fear of cancer, has led to interventions to remove the tissue most at risk for developing cancer. Angelina Jolie Pitt is the public face of a strategy to prevent *BRCA1* and/or *BRCA2* mutation carriers from developing breast or ovarian cancer by removing the breasts and ovaries before there is any evidence of the disease. Jolie Pitt told her story to the world in an op-ed published in the *New York Times* in May 2013. The following week she was on the cover of both *Time* magazine (figure 1.2) and *People* magazine. Her mother had died of ovarian cancer at fifty-nine, and Angelina herself was found to have a mutation in the *BRCA1* gene, which her physicians told her gave her a lifetime 87 percent risk of breast cancer and a 50 percent risk of ovarian cancer. As she put it, "Once I knew that this was my reality, I decided to be proactive and to minimize the risk as much as I could. I made a decision to have a preventive double mastectomy."

Angelina Jolie said she wrote about her experience in the hope that other women could benefit from it: "Cancer is still a word that strikes fear into people's hearts, producing a deep sense of powerlessness. But today it is possible to discover through a blood test whether you are highly susceptible to breast and ovarian cancer, and then take action." She wrote that her decision was "not easy" but is one "I am very happy that I made." She continued, "My chances of developing breast cancer have dropped from 87% to under 5%. I can tell my children that they don't need to fear they will lose me to breast cancer." She noted that 458,000 women worldwide die of breast cancer each year and that she wanted it to be a priority "that more women can access gene testing and lifesaving preventive treatment, whatever their means . . . and wherever they live." She concluded, "There are many women who do not know that they might be living under the shadow of cancer. It is my

1.2 Angelina Jolie on the cover of *Time*, May 27, 2013.

hope that they, too, will be able to get gene testing, and that if they have a high risk they, too, will know that they have strong options."

Americans often emulate celebrities, and *Time* magazine suggested that even women who have no medical indication for a double mastectomy might find a rationale in their genes to have this radical surgery. The chief medical officer for the American Cancer Society, Otis Brawley, for example, tells the story of a woman with no family history of breast cancer getting screened for *BRCA1* and *BRCA2* anyway. Her test revealed a mutation of "unknown significance." She had a double mastectomy. The mutation of unknown significance was later determined not to be associated with an increased risk of breast cancer. Brawley uses this story to illustrate what he terms "the pinking of America" with its overreaction to breast cancer: "We have overemphasized and scared people too much." That certainly seems to be true, and without in any way criticizing the decision Jolie Pitt made, it should be underlined that no one should make the same decision simply because she made it. Worried that others may follow her example with insufficient genetic information, the U.S. Food and Drug Administration (FDA), in late 2013, shut down the gene interpretation portion of the leading direct-to-consumer genetic testing company 23andMe. The company's TV ad included the claim, "The more you know about your DNA the more you know about yourself."

The FDA ordered the company to cease marketing its product because 23andMe could not demonstrate that its results were reliable. The company had been offering to provide genetic profiling information for $99 and give their customers, as their advertising puts it, "health reports on 254 diseases and conditions," such as heart disease, diabetes, and breast cancer. Its nationally run TV commercial, which began running in the summer of 2013 with the aim of getting one million people to purchase their product (about 500,000 had signed up when the FDA stepped in), used the following language, spoken by extremely attractive and physically fit young people: "My DNA . . . is me . . . it's like a self-portrait . . . learn hundreds of things about your health . . . change what you can, manage what you can't." Regarding breast cancer, the FDA was concerned that a false-positive result (finding the presence of the *BRCA1* or *BRCA2* gene when it is not actually there) "could lead a patient to undergo prophylactic surgery, chemoprevention, intensive screening, or other morbidity-inducing actions, while a false negative could result in failure to recognize an actual risk." This concern, however, seems far-fetched; no surgeon should operate for a genetic condition without independently confirming it.

The FDA was also worried, somewhat more plausibly, that patients who had genetic mutations that could affect the way some drugs are metabolized might change their dosage without consulting their physicians. Regarding the anticoagulant warfarin, for example, the FDA was concerned that patients taking warfarin to prevent blood clots might change the amount of the drug they were taking without consulting their physicians based on a false genotype result, and that this could lead to death or serious illness.

It also seems reasonable to conclude that the FDA acted because the agency thought it was time to regulate the entire consumer genomics industry. If so, the FDA is correct that more than just accuracy is at stake in the 23andMe debate. Other questions include whether you should have to go through your physician to obtain a genetic profile (we don't think so), who "owns" your DNA (you do) and whether you have a "right" to the information it contains (you should have), and what the role of the federal government should be in regulating DNA-information products and practices. We think regulation should be at the federal level, and that the FDA is the right agency to do it. Other commentators disagree. Eric Topol, author of *The Patient Will See You Now*, sees Angelina Jolie Pitt's public announcement as transforming her role from leading actress in action films to "playing a leading role for self-knowledge, freedom of information, and medical information ecology." In this context Topol calls the FDA's action against 23andMe unjustified paternalism.

The FDA is in talks with 23andMe, and in early 2015 the agency indicated that it was prepared to exempt genetic tests for autosomal recessive diseases that could be used by couples to determine whether or not they both carried such a gene, which would give them a one in four chance of having an affected child. The company issued a notice to their customers indicating that the FDA's approval for Bloom syndrome carrier status was the first time the FDA had authorized a direct-to-consumer genetic test.

Two years after her double mastectomy, Angelina Jolie Pitt wrote another op-ed for the *New York Times,* this time about her decision to go further and have her ovaries and fallopian tubes removed. There are no good screening tests to detect early ovarian cancer, but because of her family history, various biologic markers were being tested annually. Nothing alarming or abnormal was detected. Nonetheless, Jolie Pitt opted to have her ovaries and fallopian tubes removed—not because she had the *BRCA1* mutation, but because of family history. Three women in her family had died of cancer. Her doctors told her that preventive surgery was best done about a decade before the earliest cancer onset in her female relatives. In her words, "My mother's ovarian cancer was diagnosed when she was 49. I'm 39." With the full understanding that "it is not possible to remove all risk [of cancer]" Jolie Pitt repeated what she had said about breast cancer two years previously but this time in reference to ovarian cancer: "I know my children will never have to say, 'Mom died of ovarian cancer.' "

Two points can be underlined. First, her decision was perhaps as fully informed as is possible. For example, she understood that no guarantees came with the surgery and that there are side effects (menopause, hormone replacement therapy, physical changes, and inability to have genetic children). Second, the decision is a personal one, a part of personalized medicine. In her words, "There is more than one way to deal with any health issue. The most important thing is to learn about the options and choose what is right for you personally."

Sergey Brin's well-known story is worth recounting in this context. Sergey was born in 1973 in Moscow and came to the United States with his parents when he was six. Sergey's mother started to develop neurological problems when he was a teenager; she first experienced numbness in her hands, and later her left leg began to drag. She was evaluated at Johns Hopkins University, where she was diagnosed with Parkinson disease. During orientation for new PhD students at

Stanford, Sergey met Larry Page. Working out of a garage, Sergey and Larry started Google. The owner of the garage had a sister, Anne Wojcicki, who later became Sergey's wife (they are now separated). In 2006, Anne cofounded 23andMe, through which Sergey learned that he carried a specific mutation in a gene called *LRRK2*. This mutation was reported to give him between a 30 and 75 percent lifetime risk of developing Parkinson disease. Sergey's mother was also tested by 23andMe, and she carried the same mutation. Today, 23andMe has the largest collection in the world of DNA from people with Parkinson disease.

Sergey knew that there was accumulating evidence that lifestyle modifications might lower his risk of Parkinson disease. These include increasing exercise and coffee consumption. Sergey decided to begin exercising regularly. He goes to a pool near Google headquarters several times a week to swim. He drinks green tea on the assumption that it is caffeine intake, not coffee itself, that matters. Based on these changes, Sergey estimated that he could reduce his risk of Parkinson to 25 percent. When he includes the probable advances from Parkinson disease research in the future, including research he himself is funding, he calculated that his overall risk is only 13 percent.

It is difficult to overstate the potential of genomics to change the way we think about and treat our own medical conditions. Both health and health care will take on new genetic and genomic-driven meanings. Nevertheless, as we have emphasized, genetic information cannot explain everything in medicine, let alone in life. It is dangerous, if tempting, to take the significance of DNA information to an extreme that suggests we can both discover our human essence and precisely predict our health future by deciphering the entirety of the information encoded in the *A*s, *T*s, *G*s, and *C*s that make up our genomes. We are a product of our environment as well as our genomes. Sergey Brin understands this and knows that his genes do not alone determine his fate. Understanding that environment often plays an equal or greater role in our health future than do our genomes, he decided to reduce his risk of disease by modifying his environment through exercise and diet.

Even with significant advances in genetic testing, the information we gather can be misleading or impossible to interpret. Currently, the fastest-growing use of genomic testing is in prenatal care. Sometimes, even with access to important genomic messages, the messages are simply too complex to understand and apply to personalized medicine. For example, Sherman had a thirty-six-year-old patient, whom we'll call Donna (not her real name), who was concerned about having a child with a chromosomal disorder. Sherman did a first-trimester chorionic villus sampling (CVS), in which placental tissue is removed and tested for chromosomal abnormalities. In addition to using the traditional method of chromosome analysis, Donna volunteered to participate in a National Institutes of Health research project to have the placental sample tested using a newly developed DNA-based test called microarray analysis. Though the results showed that the fetus was not affected with any known chromosomal abnormality, the laboratory reported a DNA "variant of uncertain significance." Neither of the parents carried this variant. Although it was believed to be unlikely that the "variant" would cause significant problems in the child, Sherman could give Donna and her husband no guarantees. Instead of an answer, genetics provided a message of uncertainty, another thing to worry about. In addition to the risk of genomic tests failing to provide "actionable" information, there is also the risk that the plethora of genetic tests will simply confuse patients and their physicians and lead to anxiety and more testing.

Some physicians go overboard and provide an exhaustive list of all possible problems a variant could cause just in case lightning strikes. Their reasoning is that it's better to give too much information than too little, just in case something unanticipated goes wrong, so the physician cannot be blamed (or sued) for failing to mention the possibility. Simply increasing the amount of information conveyed, however, risks confusing the patient, losing perspective of the most important and common issues that should be considered, and engendering needless worry. It is what we have already referred to as a "personalized health threat matrix." It also illustrates an inherent paradox in modern obstetrics: in trying to make pregnancy and childbirth safer for women and increase the probability of having a healthy baby, it has been seen as necessary to perform an extensive number of tests and convey large amounts of technical, often confusing, information that predictably increase the anxiety of prospective patients.

Our final introductory example involves medical geneticist Robert C. Green of Harvard Medical School, who during the early days of direct-to-consumer genetics sent his DNA to be analyzed by all three of the new commercial companies, including 23andMe. His results indicated that he was at a lower risk of heart disease than the average person. Shortly after he got his results, he was diagnosed with serious heart disease that necessitated coronary bypass surgery. This does not, of course, mean that the results were either inaccurate or wrongly interpreted—many people with a "low" risk of heart disease in fact develop heart disease. All we have now are some population averages; we are unable to specify the risk for any given individual. The results of the famous Framingham heart study are similar. The Framingham data showed that men with a very high cholesterol level were at a much higher risk of heart attack than men with a low cholesterol level. What they could not tell, however (and we still can't), is the risk of a heart attack for a particular individual.

In 2013 Green became one of the first few hundred healthy individuals to have his entire genome sequenced. In a 2014 interview in the *Boston Globe*, Green noted that in this sequencing he carried a rare genetic mutation in the *TCOF1* gene that causes Treacher Collins syndrome. Treacher Collins affects bone development in the face and is disfiguring. Green had the mutation but insisted that he did not have the disease. This is because although the initial report described the mutation as pathogenic, he and the lab that issued the report later evaluated it as a variant of unknown significance. As reporter Carolyn Johnson put it, "Green only has to look in the mirror to know that he does not have the disorder." This is, of course, true. It is also a good example of a major point in this opening chapter: genes are not destiny, and even dominant genes for serious conditions may not express themselves. On the other hand, when we identify such a gene in an adult, we have insufficient information to counsel the person. Perhaps the best we can do for now is to say, as Green says about himself, "Most likely this is not a meaningful mutation." Finally, and most importantly, his case illustrates why it is premature to use whole-genome screening on fetuses. In Green's words, "I know this is [likely not a meaningful mutation] but imagine if you're a pregnant woman and someone reported that mutation out to you about your baby. Can you imagine?" Yes we can.

Striking a Genetic Balance

How can we maximize the good that modern medicine can do while minimizing the potential harm? This question recurs almost every time a new test, including a genomic test, is added to our medical arsenal. Couples in the near future will be faced with much more genomic information about their fetuses. Sharing these future genetic messages with a couple is imperative, but this could cause information overload and more uncertainty—and could even lead to the abortion of wanted, healthy fetuses. New methods will have to be developed to help physicians convey genomic messages to pregnant patients in a helpful and meaningful way.

Genomics will encourage some existing trends in American medicine and discourage others. American medicine, for example, mirrors four major characteristics of American society. It is wasteful, technologically-driven, individualistic, and death-denying. Restricting the use of drugs to people with genomic profiles that permit the drug to be safe and effective could actually cut down on waste. But genomics will likely reinforce the other three characteristics. The role physicians, patients and the public will play in the introduction of genomics into clinical care is uncertain. Is it reasonable to expect that our physicians will understand how to access, evaluate, and explain to us the medical implications of our genomes? How will the information coded in our genomes be read and interpreted, and by whom? Who will pay and who will want to know the content of the messages contained in our DNA? All of us have a stake in understanding genomics on at least four levels: the first is our own life and health; the second is in the lives and health of our family members; the third is in the health of our communities (including the rules under which genetic messages will be read and shared with others); and the fourth is as a member of the human species (and how we might try to modify or improve humanity).

Most centrally, we will keep asking how we can all use newly discovered and (re)interpreted genomic information in our own lives, and how we can be better prepared to deal with genomic information that could do us more harm than good. We will explore the impact of genomic messages on our potential children and our actual children. In addition to the prenatal diagnostics already in use, the future is likely to include the option of complete genome screening of newborns. How the resulting genomic messages are interpreted, and who has access to them are major public health policy questions. Should the entire genome of all newborns be sequenced and stored, or should we only screen newborns for specific genes linked to serious diseases? Should parents be able to have their children tested for genes correlated with traits such as athletic ability, perfect pitch, or eye-hand coordination as an aid to deciding what school, activity, or profession the child should be encouraged to pursue?

Some of the other topics we address in this book include the importance of genomics to fertility treatment, and the wide-ranging field of research genomics. We will examine genomic privacy in more detail and ask whether it should remain a central value, as well as whether and under what circumstances you should donate your DNA or your medical records to a research "biobank." We will consider the impact of two U.S. Supreme Court decisions on DNA. The first permits the police to take your DNA for storage if you are arrested, and the second prohibits the patenting of DNA as it occurs naturally in your body. We will also highlight proposals, which may seem like science fiction, that well-respected figures in human genomics and synthetic biology are planning to implement in the hope that they will make our lives, or the lives of our children, better, or at least longer. We begin our exploration of how genomics will affect us and our health care system by examining the meaning of personalized (genomic) medicine in the context of the American health care system.

WHEN THINKING GENOMICS, CONSIDER THESE THOUGHTS

Genomics can tell you some things about your probable medical future, but it is not destiny.

DNA is like a recipe, but you can also think of it as a book whose language is still being translated.

Genomics has already changed the way we think about ourselves and how we transmit conditions to our children.

Messages from DNA are personal and private, as private as our diaries.

There will be times you won't want (or need) to know your possible medical future; that choice should always be yours.

Genomics is evolving, and as it does, it will affect you and the way physicians interact with you in both predictable and unpredictable ways.

CHAPTER 2

Personalized (Genomic) Medicine

Enthusiasm for gene-centered medicine is contagious, and I am not immune to it. In my view, however, the fundamental issues remain. . . . Enormous amounts of new knowledge are barreling down the information highway, but they are not arriving at the doorsteps of our patients.

—Claude Lenfant (2003)

We Americans think of our health care system as the "best in the world," and that probably explains why changes in health care provoke widespread anxiety. We also see scientific progress and technological advance as good things and hope that they will ultimately make our health care, and our lives, better. Evolving genomic medicine, also known as "personalized medicine" and as "precision medicine," holds great promise for improvements in medical care, but it also will bring sometimes unwelcome changes in the way we interact with our physicians and other health care personnel. We know change is coming, but we are anxious about what it will mean for us personally and for the American health care system.

Former Obamacare adviser Ezekiel Emanuel, an expert on the Affordable Care Act, begins his book on the future of American medicine, *Reinventing American Healthcare*, with the story of Erin and Justin and their two daughters. Justin dies tragically in a skiing accident. A few years later, Erin gets very sick after a trip to Mexico. Antibiotics fails to cure her, and eventually she is diagnosed with a strangulated colon that necessitates emergency surgery. The surgery discloses cancer, which the surgeon is pretty sure he has removed, so chemotherapy may not be needed. The surgeon, however, recommends a genetic test to see if she has Lynch syndrome, a genetic disease predisposing one to early colon cancer (Erin is forty-eight). The genetic test does not find Lynch syndrome, but specialists recommend additional genetic testing. Emanuel tells this story to illustrate how much medical care costs, the incredibly complicated insurance system we have, the problem of defining a pre-existing condition, and the reluctance of many health care providers to see uninsured patients, even critically ill ones. The Affordable Care Act will make payment easier for many Americans. For us, however, the point of this story is that genetics is still seen as an afterthought in medical care, and that as cutting edge as it is, Emanuel never mentions genetics again in his book about our current health care system.

Emanuel's failure to describe the genomic medicine of the future is understandable: it's still too early to tell how this evolving area will play out, or even how it should be described. President Obama, for example, called it "precision medicine" in his 2015 State of the Union address. And what is still widely referred to as "personalized" medicine may actually be more impersonal than today's medicine. Personalized, genomic medicine requires electronic health records, and you may already find your physician interacting more with a computer than with you during office visits. You will need a basic understanding of the new language of genomic medicine, and the most likely ways it could impact your care, to be able to navigate the evolving genomics-enabled health care system. Personalized medicine has been one of the most hyped "revolutions" in modern medicine. In this chapter we put the current state of personalized medicine into perspective and suggest ways you can use genomic information in everyday health care for you and your family. Our goal in arming you with this information is to help make you a better advocate for your own health and that of your family. We begin with the rhetorical shift from personal to personalized medicine.

Forty years ago, a grandfatherly physician from Kentucky came to speak to Sherman's medical school class about his memorable patients. He also offered the medical students two pieces of advice. First, after entering a patient's hospital room, always sit down, even if the only place to sit is the patient's bed. Second, always touch the patient. (Today the physician would have to first wash her hands.) It doesn't matter what you do: take his blood pressure, feel his abdomen, look into his ears, or just put your hand on his forehead to feel for a fever. The critical element is physical contact.

This is, we think, the kind of "personalized" medicine most of us hope for, if not expect, from our doctor when we are sick or hurt. Of course, we presume that our doctor is competent and skilled, and has access to the latest medical technology, but that's not enough. We want a personal relationship with our doctor; we want high tech, but we also want "high touch." None of us want "cookbook medicine"; we are all unique, and we rightly expect our doctor to take into account our medical history, together with the results of physical exams and medical tests, to develop an individual treatment plan that is optimal for us.

Several terms have been used interchangeably with *personalized medicine,* including *precision medicine, stratified medicine, targeted medicine,* and *genomic medicine.* The best definition of personalized medicine so far comes from the National Academy of Sciences: "The use of genomic, epigenomic, exposure, and other data to define individual patterns of disease, potentially leading to better individual treatment." The last clause is not really part of the definition, but rather expresses the hope that genomics will improve patient outcomes. *Personalized medicine* is a misleading label for the use of genomics in health care because it implies that you are your genes, and by treating your genes physicians are treating you. This is far too limiting a view of personalized medicine. As physician Marshall Chin put it in the context of improving patient care: "Truly patient-centered care is individualized care by clinicians who appreciate [that] patients' beliefs, behaviors, social and economic challenges, and environments influence their health outcomes." Chin continues that the "best care" spans years and includes inpatient, outpatient, and self-care that is "tailored" to the individual needs of each patient.

All this helps explain why we prefer the term *genomic medicine* to *personalized medicine:* it more accurately states what it is, using genomic information (information from our genome) to help guide health care. To us, and most likely to you as well, *personalized medicine* conjures a different image, a more holistic medicine like that practiced by the grandfatherly Kentucky physician. Nonetheless, it is difficult and confusing to try to change labels once they have become part of our medical vocabulary. We use both *personalized medicine* and *genomic medicine* in this book (we don't think the term *precision medicine* will have a long life), depending on which aspect of personalized genomic medicine is being emphasized. We encourage you to think *genomics* whenever you see *personalized.*

Genomic medicine does not always require sophisticated and costly genomic technologies. Your family history is the simplest, most straightforward, and least expensive way for you to assess the risks of inherited medical conditions. For all these reasons, it is important for your health and health care that you know your family history. Popularly called a family "tree," it has its origins in genealogy, identifying your relatives by their names and recording your family lineage. Observations that certain maladies run in families have been documented throughout history, and most people believe they will develop the diseases their parents had. This is not always true, but understanding family relationships is the beginning of genomic medicine.

The Genetic Family History

Referring to direct-to-consumer genetic testing, one medical expert at the U.S. General Accounting Office went so far as to tell Congress that "the most accurate way for these [direct-to-consumer DNA profiling] companies to predict disease risks would be for them to charge consumers $500 for DNA and family-medical-history information, throw out the DNA, and then make predictions based solely on the family history information." This is an exaggeration, but it properly emphasizes both the power of a family history and the insufficiency of existing data to support genetic testing inferences.

Taking a genetic family history involves systematically inquiring about your health status and ethnic origin, first-degree relatives (siblings, parents, and offspring), second-degree relatives (uncles, aunts, nephews, and grandparents), and third-degree relatives (first cousins). The birth dates for living family members, and age and cause of death for deceased family members, should be recorded. It is especially important to check if there is any relationship by descent from a common ancestor (called consanguinity or "blood relative"). It may be necessary to obtain medical records (including laboratory test results, imaging studies, and pathology reports) to confirm directly the diagnosis of a relevant disorder in a family member. Untoward pregnancy outcomes (for example, miscarriages, stillbirths, infants born with congenital anomalies) should be ascertained. Exposure to drugs (including prescription, nonprescription, and "street drugs") and toxic chemicals should be determined. The amount and duration of alcohol and cigarette smoking should be evaluated. Depending on the situation, particular emphasis should be placed on obtaining detailed information about the health concern in question (for example, cancer, prenatal diagnosis, or dementia).

Genetic professionals usually assess the risks of inherited medical conditions by performing a *pedigree analysis*. A pedigree is a diagram of family relationships that uses standardized symbols to represent people and lines to represent genetic relationships. Pedigrees ideally show at least three generations, mark individuals affected with a specific diagnosis, and indicate ethnic ancestry (figure 2.1). This display can be helpful in identifying how genetic disorders are inherited. A pedigree can be easily drawn by hand and scanned into the medical record. In practice, most health professionals use questionnaires, often with checklists. There is an interactive software program that allows you to record your family history on a computer. One version, called *My Family Health Portrait: A Tool from the Surgeon General,* is available free online. There are also inexpensive apps for making your family pedigree.

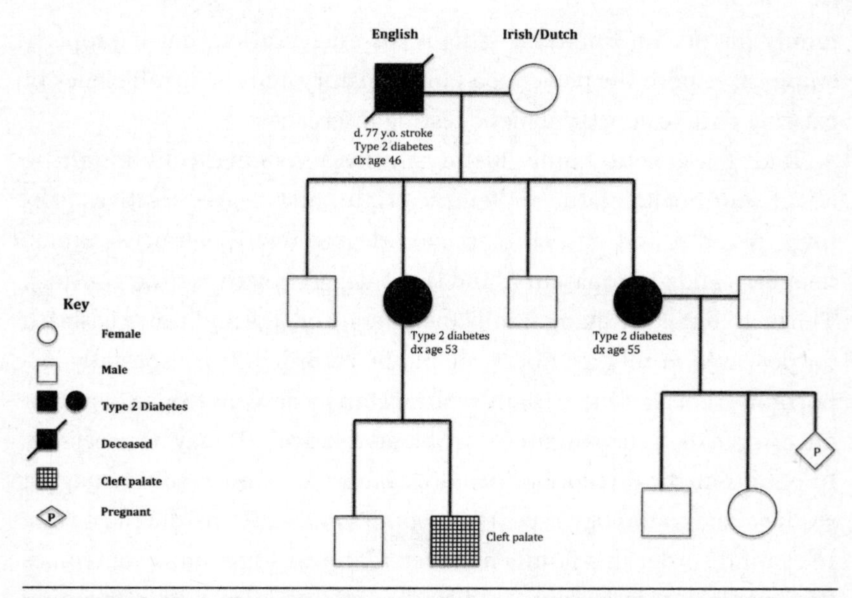

2.1 Example of a pedigree showing type 2 diabetes and cleft palate. Abbreviations: dx, diagnosed; y.o., years old; P, pregnant. Personal image.

Compared to depending only on notes in the medical record, using a family history and pedigree tool increases the likelihood of identifying a person at risk for an inherited medical condition by about 20 percent. In fact, family history and other conventional tests (for instance, cholesterol testing and blood pressure) outperform genomic-based risk predictions for cardiovascular disease. When you hear about how genomic testing discoveries are giving us valuable information about the underpinnings of disease, including the promise that these discoveries will soon lead to long-awaited new treatments, you should be skeptical, like a good scientist. For many common diseases, such as type 2 diabetes and certain cancers, the initial hype about breakthroughs has not translated into meaningful changes in everyday health care.

As mentioned in chapter 1, the overpromising of direct-to-consumer company 23andMe led the FDA to warn the company to stop advertising its personalized genetic tests in late 2013. The FDA was concerned that the company's national advertising campaign was making promises it could not keep, including that customers could "learn hundreds of things about your health," for instance that you "might have an increased risk of heart disease, arthritis, gallstones, [or] hemochromatosis." The company's commercials also suggested that the genetic information they could derive from analysis of your DNA was sufficient to permit you to change your life and your health: "Change what you can, manage what you can't." The FDA was correct, we think, in effectively shutting down 23andMe's disease-risk profiling service for now. Family history remains more informative than limited genetic screening—and family history can help determine whether genetic screening makes sense (for instance, in the case of the breast cancer mutations).

Physicians seem to agree, at least based on an unscientific poll of physicians conducted by the *New England Journal of Medicine* in late 2014. Readers were asked whether they would recommend genomic screening for a hypothetical healthy, asymptomatic forty-five-year-old male, Jim Mathis. Mathis told his internist he was concerned about his risk of cancer after completing a family tree at a genealogy workshop, where he learned that three relatives had cancer (breast, ovarian, and prostate). Of the approximately one thousand physicians responding, 40 percent would not perform any genomic testing, and most of those who chose to test would limit testing to cancer genes. Only 12 percent would do the entire genome. The summary of the results concluded by focusing on the patient rather than the physician respondents: "Many patients, like Mathis, are curious and will actively seek knowledge about their genetic risk."

In the aftermath of the FDA's action, 23andMe limited its services to building family trees, advertising that you can use its services to "build your family tree and enhance your experience with relatives." You can also "learn what percent of your DNA is from populations around the world" and "contact relatives across continents, or across the street." Other companies also offer family-tree-building services, including Ancestry, WikiTree, and Geni. *Esquire* editor-at-large A. J. Jacobs has noted that by using the services of Geni, whose database currently has 75 million "relatives" from 160 countries, he was informed that he is related to more than 80,000 other humans. In his words, "My newfound kin include the actress and lifestyle guru Gwyneth Paltrow, a mere

17 steps away, and the jazz great Quincy Jones, a mere 22. . . . These folks have no clue who I am." Jacobs envisions the near future when our family trees will include "all seven billion humans on earth"—since we are all genetically cousins—and asks, "If everyone is related, what does the concept of family even mean?"

It's a good question, since humans are all genetically related at some level. One outcome of all this consumer genealogy research based on genetics could be a restructuring of groups that we identify with, away from groupings based primarily on geography or religion and toward groups based on genetic similarities. People want to know more than just which continents and what percentages of each are represented in their DNA. Medical researchers want to know more than that too. They want to know which people share which genes, so they can study how these genes are expressed in various environments. Members of genetically similar "groups" in the not too distant future may themselves form Facebook communities based on genetic similarities, as much out of curiosity as an exercise in family building. In short, we have so far used surrogate indicators like kinship, religion, and geographical origin to suggest what genes a person may share with others. Soon we will be able to sequence the entire genome of an individual, who can then directly compare his or her genome with that of others. Then, to the extent we want to, we can form new group identities based directly on our unique genomes and how they compare and overlap with the genomes of others.

Impersonal Medicine

It will be a while before genomic analysis can provide more useful information than the family tree, and in the meantime American physicians seem to be concentrating more on their new technologies than on us—the opposite of the personal medicine most of us want. One of the most famous physicians in the world, Arnold Relman, the late editor of the *New England Journal of Medicine,* describes the technically brilliant lifesaving treatment he got at Massachusetts General Hospital after he broke his neck in a fall in 2013. There were no personal conversations with physicians, who spent most of their time "with their computers" and examining "copious reports of the data from tests and monitoring devices." Relman notes, "What personal care hospitalized patients now get is mostly from nurses." Nurses, of course, have always been the 24/7 caregivers in hospitals, but have they now almost entirely taken over the job of communicating with patients as people? Can genomics help reverse the trend of physicians in hospitals concentrating more on the computer and the patient's electronic medical record than on the patient him- or herself, or will genomics make medical care even less patient centered? We think that genomics, at least in the immediate future, will encourage the trend toward more impersonal medicine (rather than personal medicine), and only strong patient and family voices are likely to reverse this trend.

When you are admitted to the hospital, you can expect "Big Medicine." Many economists, and even some physicians, think that giant private corporations, which see maximization of profits as their primary goal, should be seen as role models for health care. We strongly disagree. We should be maximizing the health of Americans, not the profits of their caregivers. Surgeon-author Atul Gawande, for example, has suggested that our health care system has much to learn from the Cheesecake Factory, whose 160 restaurants serve more than 80 million people a year. Gawande believes that one key to the success of the Cheesecake Factory is its size, that economies of scale permit it to provide food and service of greater quality at lower cost. Size gives their restaurants buying power, allows them to centralize common functions, and enables them to adopt and diffuse innovations more quickly than do small independent operations. Our health care system has a similar goal: trying to deliver a range of services to millions of people at a reasonable cost and with a consistent level of quality. Yet American medicine has been largely unsuccessful in meeting these goals, producing instead skyrocketing costs, mediocre service, and unreliable quality.

How does genomic medicine fit into the Cheesecake Factory model? A key to the success of the Cheesecake Factory is standardization: making sure each dish looks and tastes the same time after time. This means that there can be no room for varying the recipe. The Cheesecake

Factory has thirty types of cheesecakes on its menu. What if there were 13,360 kinds of cheesecake? In a discussion at Harvard Business School, Gawande noted that "healthcare is incredibly complex simply because of the myriad ways—13,360 and counting—that the human body can fail." He observed that the delivery of health care is "arguably the delivery of 13,360 service lines, town by town, to anyone who needs care."

Now think about what would happen at the Cheesecake Factory if each cheesecake had 3 billion ingredients, just as the human genome has 3 billion base pairs, and no two cheesecakes had the same ingredients, just as no two individuals have exactly the same genome. No two cheesecakes would look and taste exactly the same, just as no two people are exactly the same. The point is that there is no standard genome; personalized genomic medicine is directly dependent on DNA variation. As genomic medicine is increasingly used in health care, the focus will be on your genetic uniqueness as spelled out by your DNA. Genomic medicine will be adding a lot more information to a system already overloaded by more information than any other industry. This flood of genomic information will require sophisticated bioinformatics and computer-based translation in the doctor-patient relationship. We think genomic medicine will eventually make for better medical care, but it will *not* make medical care less expensive or more efficient, especially in the short run.

We can accept the Cheesecake Factory as a good model for the hospital cafeteria without applying it to the intensive care unit or the operating room. The real challenge for our health care system will be how to adopt many of the good features of megacompanies like Walmart and Amazon, such as emphasis on quality measures, cost efficiencies, and customer satisfaction, without losing sight of the fact that every patient really is unique (and not just genetically) and deserves to be treated and respected as an individual. We remember, for example, the hospital administrator who, in the midst of the 1990s fad of managed care, said he didn't see any reason why a hospital could not be run like a ball bearing factory. And he would have been correct if ball bearings got sick, suffered, died, and had families that cared about their welfare. There are, nonetheless, routine steps that should be taken with every patient. Gawande is on firm ground here when he proposed that surgeons adopt checklists, like those used by pilots before flight, to use before they begin surgery. Having all surgeons check to make sure they are doing the right operation on the right patient at the right site, for example, makes perfect sense and should be mandatory.

Other experts, like author Robin Cook and physician Eric Topol, see a future in which your smart phone becomes an "avatar physician." It will monitor your heart rate, respiration, and so on, and use this data to determine if you are having a heart attack and call 911. It may also be able to predict heart attacks or other conditions and warn you about them. Topol also envisions everyone having their DNA sequenced and made part of their electronic health record so that clinicians can "match individual with treatments." The goal is not, Topol assures us, to replace your physician with your smart phone and computer, but, with your smart phone doing the routine monitoring and testing, "your flesh and blood primary care physicians will have more time to talk to you when you do need to see them." We're not convinced this is medicine's future, but we concede that self-monitoring devices are proliferating. You may even want to try one yourself since, for at least some people, they can help provide the motivation to increase exercise or change to a healthier diet. We strongly agree with Topol that your medical and genomic information is yours and you should have complete and direct access to it.

Genetically Isolated Populations

Studying isolated populations has produced much of what we know about genetics and heredity. Understanding why will help you understand much of the genetics in this book. It has long been known that in animal communities inbreeding leads to problems and outbreeding leads to "vigor." The same can be said for human populations. You and your first cousin can expect one-eighth of your genes to be identical by descent from the same ancestor. If individuals in a small community marry only among themselves, over generations everyone becomes genetically related. From a shared gene perspective, all marriages are among cousins. An examination of the family trees (pedigrees) of couples living in small, isolated communities reveals that their roots appear to twist back into themselves multiple times. This results in "genetic bottlenecks" within the population. For example, grandparents might be related to each other in multiple ways; for a couple marrying, each grandparent may be related to the grandparents on the other side. It's easy to see how, generation after generation, the likelihood of a couple sharing the same ancestral genes steadily increases.

This inbreeding effect has led some autosomal recessive diseases (diseases expressed only if the person has gotten the same mutated gene from both parents and is thus homozygous for the mutation) to be more common in particular populations. If a couple shares a mutation in the same gene, there is a one-in-four chance of having a child affected with the disease coded for by that gene. Examples include Tay-Sachs disease among Ashkenazi Jews; beta-thalassemia among individuals of Mediterranean descent; cystic fibrosis among individuals of northern European descent; sickle cell anemia among individuals of African descent; and alpha-thalassemia among individuals of Asian descent. A number of unique isolated populations have been studied extensively in genetic research. We describe three: the Mormons, the Amish, and the Hutterites.

Geneticist Ray Gesteland of the University of Utah believes that "more human disease genes have been discovered in Utah than in any other place in the world." This is because of three large data banks: the Utah Population Data Base, the Utah Cancer Registry, and more than 185,000 genealogical records from the Mormon Family History Library. For more than three decades, researchers at the University of Utah have collaborated with the Mormon Church in numerous genetic investigations, particularly in cancer research. Mark Skonlick, a geneticist at the University of Utah, for example, used Mormon medical records and pedigree information to help establish Myriad Genetics in 1990. The company later identified and isolated two breast-ovarian cancer genes, *BRCA1* and *BRCA2*.

Another religious group, the Amish, have 275,000 members in more than thirty states. The Amish avoid most modern technology and prohibit or limit \use of telephones, television, and automobiles. Genetic studies on the Amish began in the early 1960s, conducted by Victor McKusick of Johns Hopkins University, after David Krusen observed that a form of dwarfism (originally described as achondroplasia but later diagnosed as Ellis–van Creveld syndrome) was frequent among the Amish. Cartilage hair hypoplasia (CHH), an autosomal recessive form of dwarfism associated with immunodeficiency and higher risk of developing lymphomas and leukemias, was first characterized in the Amish. McKusick found advantages in doing genetic studies with the Amish: their genealogic records are extensive; they are interested in the causes of illness; there is a high coefficient of inbreeding; they are relatively immobile; and they keep children with genetic abnormalities at home rather than institutionalizing them. Nor do the Amish object to modern medicine. The Clinic for Special Children in Strasburg, Pennsylvania, may be the only clinic in the world where you will find DNA sequencers inside and hitching posts outside.

Earlier in his career, Sherman was involved in a research project at Northwestern University studying the Hutterites (figure 2.2). The Hutterites first settled in what is now South Dakota, establishing three communal farms (colonies). When a colony reached a certain size, usually around 150 people, it would split. By drawing lots, half the families would remain in the old colony and half would move to the new one. Today there are more than 40,000 Hutterites living in almost four hundred colonies. Hutterite colonies often have a one-room schoolhouse where the children are taught secular subjects for half the day and religion the other half. Hutterites follow a communal lifestyle with shared goods, eating in a dining hall where the men and women sit apart. Children eat separately.

The Hutterite communal lifestyle is ideal for human genetic studies: they are exposed to similar environments, eating an old-style Germanic diet, with no smoking, no birth control usage (the average Hutterite couple has five children), and only occasional alcohol consumption. Most importantly, virtually all current Hutterites are descendants of the original founding population. The colony preachers keep meticulous

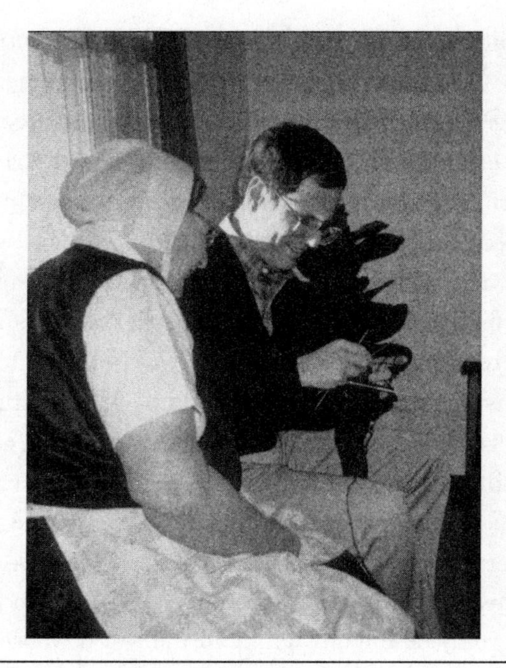

2.2 Sherman with the Hutterites. Personal photo.

genealogy records and have recorded the marriages, births, and deaths of colony members for hundreds of years.

Over the years, Sherman drew blood samples from hundreds of Hutterite men, women, and children. Coincidentally, Sherman's last name, Elias, is a common Hutterite name. Members of the visited colonies often forewarned new colonies that a doctor named Elias would be coming. The Hutterites would frequently ask if he had Hutterite ancestry, and when he told them he didn't, they told him he just needed to look harder. The extensive pedigrees and medical histories of the Hutterites, linked with the studies performed on their blood samples, permitted invaluable insights into the genetics of such conditions as diabetes, asthma, breast cancer, and repeated miscarriages.

Two main points from the genetic research done on these populations merit emphasis. The first is that a key element of research in genomic medicine is obtaining an accurate, extensive family history. This is why unique populations such as the Mormons, Amish, and Hutterites are of such interest to genomic researchers. With or without genomic information, family history (including ethnicity) can provide highly valuable information for you and your physician that can directly affect your health care. The second is that genetic conditions can be inferred from family histories and later confirmed by genetic analysis. As whole-genome screening becomes more widely used, however, genomes can be (and will be) compared directly, without the need to look for kinship relationships. For example, in 2014 another "isolated" population, the indigenous population of Saudi Arabia, embarked on a project to do whole-genome sequencing on 100,000 members of this population to determine to what extent marriages among first cousins resulted in genetic abnormalities in their children.

Electronic Health Records

Genomics will be a key element in the development of evidence-based medicine because individuals with varying genomes will respond differently to the same drugs. In *A Study in Scarlet*, Sherlock Holmes observes to Dr. Watson, "It is a capital mistake to theorize before you have all the evidence. It biases the judgment." Unfortunately, for a large part of medical practice, there is surprisingly little evidence for what works, or doesn't work, and what is best for individual patient care. Recognizing this, about twenty years ago a small group of medical researchers and educators began advocating for what is now known as evidence-based medicine (EBM). Simply put, EBM "is the integration of best research evidence with clinical expertise and patient values." The goal is to improve patient care.

In comparative-effectiveness research (CER), patient outcomes of one approach for managing a disease are compared to other approaches—for example, comparing the effectiveness (in terms of a defined outcome, such as survival rates) of two or more drugs for the same disease. CER is an effort to find out what works best, but it can also be an effort to save money by not paying for useless or marginal care. This approach can help us determine if we are getting the biggest bang for our health care dollars. Collecting data on treatment outcomes is essential to improving the quality of medical care, and we think you should enthusiastically support this effort. It is widely recognized that much, if not most, medical care today can accurately be described as "gray" care—care that is routine, does little if no harm, but does not do much, if any, good. The goal with genomics is to introduce it into clinical medicine in a way that both improves patient outcomes and is cost-effective.

Some treatments and screening tests can do more harm than good. Prostate-specific antigen (PSA) screening for prostate cancer is a good example. In 2012 the U.S. Preventive Service Task Force made a highly controversial recommendation that affects nearly 45 million men annually. After decades of routine use in clinical practice, the task force recommended that we abandon screening for prostate cancer with PSA blood testing because it causes more harm (for example, unnecessary biopsies, overdiagnosis, overtreatment, and complications, such as urinary incontinence, erectile dysfunction, and bowel dysfunction) than benefits (that is, potential survival advantages from treating prostate cancer). Even the suggestion that men give up annual PSA screening met with stiff resistance, particularly from the American Urological Association and prostate cancer advocacy groups, who have a financial stake in not changing current payment policies. Understandably, this conflict has led to considerable confusion, frustration, and even anger among men. Currently we are left with a middle-of-the road "shared decision process" that satisfies no one. To share decision making with your physician is fine, but the decision is not an informed one without the relevant evidence. For more on screening tests, see Appendix B.

How does taking a simple family history stack up in this era of EBM and CER? And how does a family history compare with genomic testing? In both EBM and CER, groups of patients, not individual patients, are analyzed to compare the effectiveness of alternative medical treatments. This may seem to be at odds with the fundamental tenets of genomic medicine, where the focus is on the individual patient's unique and specific disease characteristics, coexisting conditions, genetic factors, risk factors, and personal values and preferences. Mostly, however, genomics will be used not to treat you uniquely but to treat you the same way as other people with your relevant genetic traits. In this respect, "personalized" medicine can be seen as genome-based group or sector medicine ("stratified" medicine), with the group or sector of the population based on genetic commonalities, similar to those identified in the four genetically isolated groups we discussed.

Proponents of genomic testing, particularly direct-to-consumer companies, have suggested that personalized medicine will encourage people to positively modify their lifestyles, especially making changes in diet and exercise. However, the same can be said for taking a family history. One randomized controlled trial showed that systematically screening for family history and tailoring prevention messages can be effective in improving health behaviors by increasing physical activity and fruit and vegetable intake. If all you get from having your DNA tested is a recommendation that you should exercise more and eat a healthier diet, taking a family history is much more cost effective. Of course, it's another matter if genomic testing (which can be indicated based on a complete family history) can establish that you carry a particular deleterious gene, such as the *BRCA1* or *BRCA2* mutation. This mutation puts you at substantial increased risk of developing breast cancer, ovarian cancer, or both, and its identification could lead you to consider proven, effective preventive measures, including the drastic step of having a bilateral prophylactic mastectomy and removal of the ovaries, as illustrated by Angelina Jolie Pitt (discussed in chapter 1).

As we have suggested, genomic information will become much more useful to you and your physician when it is integrated into your medical record. This will not be realistic until your medical record is kept digitally. Most hospitals either have converted or are in the process of converting their medical records (now usually called a "health record") from paper to computer. Electronic health records (EHR) are a digital version of the old paper records. In addition to being more legible than handwriting (which can prevent medical errors), EHRs have many advantages over paper records. EHR allows computer storage of medical records, laboratory tests, digitalized images (for example, X-rays, ultrasounds, and MRIs), electrocardiograms, fetal heart rate tracings, and so forth. This requires far less space and lower associated costs compared to storing hard copies. EHRs allow rapid retrieval, searches, and collating of medical information. They permit your doctor to easily chart information about you over time (for instance, blood pressure and weight) and track when you are due for screening tests (such as a mammogram or colonoscopy). In a hospital, EHRs can also be used to track infections, as well as trends in admissions and discharges. In fact, your records are designed to be accessed by everyone involved in your health care, including you. Because of incompatible systems, however, the information in EHRs doesn't travel easily out of the practice or hospital.

Transition to the use of EHRs is well under way. The adoption of EHRs by the American health system will expedite the introduction of genomic medicine into everyday practice. The likely future interaction of the EHR and genomic information is illustrated by the medical treatment of a hypothetical patient we'll call Olivia. Her treatment can be seen as a best-case scenario of the future of genomic medicine. Olivia's case can be usefully contrasted to the case of Erin, the patient whose story we related at the beginning of this chapter, who had no unified health record, electronic or otherwise.

Olivia came to the United States with her family from China when she was four years old. Shortly after she turned seventeen, she suffered a grand mal (generalized tonic-clonic) seizure. Using an electronic referral system, her primary care physician sent Olivia (and her EHR) to a neurologist, whom we'll call Dr. Good. At a university medical center, Good performed a number of tests, including an electroencephalogram to measure electrical activity in the brain. Good decided that the best drug for her condition would be carbamazepine (Tegretol®) and entered the prescription in Olivia's EHR. The software performed a "carbamazepine pharmacogenetic adverse event algorithm," which required Good to enter a number of mandatory "data elements" for the "decision model" before the pharmacy could accept the prescription.

A history of allergy to carbamazepine was queried, as well as any potential interactions with other drugs Olivia was taking. To determine the correct dose, the computer needed the patient's height and weight. The possibility of pregnancy was important because fetal exposure to carbamazepine is associated with an increased risk of birth defects (including head or facial deformities, spina bifida, and heart defects). Olivia was a young female, so the computer prompted Good to order a pregnancy test. Personalized genomic medicine also came directly into play. Olivia's Asian heritage produced a prompt to check her HLA-B*1502 status, which increases the risk of developing a potentially deadly blistering skin reaction to carbamazepine called the Steven-Johnson syndrome. Because she had this genotype, the computer suggested alternative drugs, and Good ordered one of them. Olivia's treatment illustrates how the EHR will advance genomic medicine by embedding a patient's genomic information into clinical decision making. Decisions affected include disease risk assessment, accurate disease diagnosis and subtyping, drug therapy and dose selection, assessment for adverse drug reaction, and family planning. The EHR has the potential to bring a tremendous amount of "personalized" information directly into the doctor-patient relationship.

Integration of genomics into medical care is evolving and will be gradual. It is part of overall efforts to improve patient outcomes and confront ever-increasing health care costs. Increasing the quality of medical care, however, does not automatically decrease its overall costs. Medical care is not like making cheesecake. Few of us want to go to Walmart for our physicians and hospitals, looking for the cheapest health care possible. It is likely that more information will lead to more expense. A study of 30,000 patient visits to more than 1,000 office-based physicians, for example, found that use of EHRs resulted in ordering additional imaging and laboratory testing.

The genomic information that will be included in your EHR will evolve. At first, as with Olivia, information about particular genes will be entered. Eventually, we think likely within a decade, we will transition to entering your whole-genome sequence, all 3 billion base pairs. There is still a question as to whether the "raw" genetic data will be entered into the EHR itself or housed in a cloud-based or other storage system separate from the physician and hospital records. Regardless, as genomic information is used more widely by clinicians, systems will have to be in place to regularly "interrogate" your genome. Some of this information will be of critical importance to your health care, but much, if not most of it, will be useless, and some will actually cause confusion and even potential harm. The challenge, of course, is to maximize the useful genomic information and minimize the marginal, useless, or even potentially harmful genomic information. As the director of Duke's Center for Human Genome Variation accurately predicted in 2011, "Within the next few years our ability to identify pathogenic and potentially pathogenic mutations—as well as huge numbers of mutations that no one can vouch for as dangerous or safe—will almost certainly outstrip our ability to act on the information." Nonetheless, we expect that within a decade or less everyone (at least all adults) will have direct access to their entire medical records by personal computer, including their genome. This will put a lot of discretion in your hands, and you will have to decide how much you want to know about your genome. Your answer will likely be highly influenced by what actions, if any, you can take to modify the effects of your genes on your health. Answering this last question is what most contemporary genomics research is about.

In the chapters that follow, we also address the hype of genomic medicine and how it has been oversold in large part to maintain public support (and more importantly, federal NIH funding), as well as by the promise of financial bonanzas for private industry. This doesn't mean that genomic medicine is not going to be useful. Just the opposite: we believe it is evolving to be a major force in medical practice. It will likely affect virtually every aspect of how we diagnose, treat, and prevent illnesses and medical conditions—everything from diabetes, to infections, to infertility, to prenatal care, to dementia, to cancer. In other words, we think genomics will follow the path of most transformational technologies, whose impact is overestimated in the short run but underestimated in the long run.

WHEN THINKING PERSONALIZED (GENOMIC) INFORMATION, CONSIDER THESE THOUGHTS

Whenever you see "personalized" or "precision" think genomics.

Your personal family history can often tell you more than your genome.

Your family history can help determine whether to have genomic testing.

Much of what we know about genomics comes from the study of isolated populations.

Genomics will not be used routinely in medicine until your genomic sequence can be stored on, or linked to, your electronic health record.

CHAPTER 3

Nature, Nurture, and the Microbiome

Everything is environmental until you convince me that it is genetic.

—Barry Marshall (2014)

Shortly after President Obama announced his 2015 genome initiative to save lives and cure cancer, physician-*New York Times* essayist Abigail Zuger wrote that her own experience with patients is that the patient's environment is much more important than the patient's genetics. She described her admittedly extreme patient, Barbara, as a smart, homeless, alcoholic who suffered from "medical problems as predictable as spokes on a wheel: bad heart, terrible liver, crumbling hips, gummed up lungs, AIDS from a brief foray into injectable drugs." Always on the verge of changing her life, Barbara nonetheless often missed her medical appointments. Zuger agrees that it may turn out that many of Barbara's problems are related to her "genetic predispositions," but suggests that more knowledge about her genome is unlikely to help her much. Zuger suggests that the federal money the president is proposing for more genomic studies could be better spent on her Barbara's other needs, "like supportive housing with on-site counselors and addiction services."

Yes, Barbara is an "extreme example." Nonetheless, she shows starkly that "genes are seldom the whole story behind illness . . ." As Barbara's story highlights, environment and lifestyle play at least as important a role. Genomics will not solve homelessness, addiction or poverty. Public health will continue to be more important than medicine in actually making the lives and health of populations better because of the importance of our environment on our health. Genomics could, nonetheless, help some of us. This is primarily because we can often control at least some of the environmental factors that influence the expression of our genes. For example, if you are genetically predisposed to being overweight and having high blood pressure, eating a healthier diet and getting regular exercise could reduce your risk of diabetes.

In this chapter, we address the interactions of our genomes and our environment by examining the genetics of cloning, the influence of the uterine environment on the fetus, the influence of the environment and genetics on diabetes, and the interaction of both genetics and the environment with our microbiomes. These topics are all interrelated, and the goal is to demonstrate that our genomes alone do not determine our health; genes interact with each other, and with our environment and microbiomes, in ways that can be influenced by our own actions.

Twins and Genetic Research

Cloning is a method of producing genetically identical animals (twins) born years apart—a technique that can be used to observe gene-environment interactions and effects. The most famous cloned animal and perhaps the most famous animal in the world was Dolly the sheep, a Finn-Dorset ewe (figure 3.1). Dolly was born in the laboratory of Ian Wilmut and Keith Campbell in Scotland in 1996. Almost instantly, she became an international celebrity. Her fame was based on her shocking novelty: she was the first mammal whose genetics was derived almost entirely from a body cell or "somatic cell" (to differentiate it from a "germ cell," an egg or sperm) removed from an adult—a feat most scientists thought was simply impossible. She was gestated from an embryo produced by fusing a cultured adult somatic cell with an egg from which the nucleus had been removed. Because her nuclear DNA was identical to that of her genetic mother—the ewe from whom the somatic cell had been removed—Dolly was a genetic duplicate, a later-born identical twin. She is usually referred to simply as a clone.

Genetics did more than label Dolly a twin; genetics was also used to identify her mother—the dead ewe that was the source of the somatic cell used to construct the embryo. Dolly's two other mothers—the egg donor and the ewe that gestated and gave birth to her—were never seriously considered as either exclusive or joint mothers to Dolly, even though they completely determined her environment from fertilization to birth, and the egg donor also supplied mitochondrial DNA.

The international debate Dolly sparked was and is primarily about how important genetics is to personal identity and what impact environmental factors can have on genetically identical animals (or people). Is there, for example, any reason we should not use the cloning technique on humans to try to create babies that are genetically identical to an existing person? George joined Ian Wilmut and others to testify

3.1 George with Dolly at the National Museum of Scotland, Edinburgh, 2012. Personal photo.

on this subject before a U.S. Senate panel in early 1997. George and Ian (who was the star of the hearing) agreed that human cloning should not be pursued. On the flight to Washington for the hearing, George sat next to political analyst James Carville, who opined (accurately) that human reproductive cloning would not be attractive to the public because it has "no payoff." Parents want their children to have better lives than they had, not to simply "duplicate" their lives. It is also likely that the later-born "twin" (the clone) would be seen as a copy of a more valuable existing person (the "original"), and that the twin's value (to herself and others) would be lessened because her future life is seen as dependent upon her genome rather than on anything she herself does.

A contrary argument is that cloning should be categorized as just another method of human reproduction, and infertile couples should be able to employ cloning to have genetically related children (at least if there is no other way for them to have genetically related children). The paradoxical presumption behind this argument seems to be that the only children who count as "your own children" are those with whom you or your spouse have a genetic tie, that the genetic tie is (much) more important than the (environmental) ties of rearing a child. Although genetically identical to her genetic mother, Dolly had different markings on her coat and suffered from different illnesses, including arthritis. She also died young, at six years, from a contagious lung disease she caught from another sheep—proof that genetics alone is not destiny; our destiny is also heavily influenced by our environment.

In 2012, two scientists shared the Nobel Prize in Medicine for their work on reprogramming cells to be used for cloning. Cloning on the cellular level to produce tissues for medical treatment is known as "therapeutic cloning" (more accurately, gene transfer experiments) and is to be contrasted with cloning to make an entire animal, which is asexual "reproductive cloning." The great advantage of therapeutic cloning to make replacement tissue (instead of using tissue donated from another person) is that any tissue made from a patient's cells carries the patient's own genome, and so will not be rejected. In the spring of 2014, two groups of scientists used this technique to produce human embryonic stem cells (from fusing skin cells of two adult men in one case, and cells from an adult female in the other, with an enucleated human egg). The attempts required hundreds of human eggs, making the procedure unsuited for clinical use at this time, but this was a major milestone in gene transfer experimentation.

Experimentation on cells, at least prior to the human embryo stage, is seldom controversial. Animal and human experimentation, however, can raise complex ethical problems. Even though Dolly, for example, was a well-treated celebrity, she was confined to a pen during her short life. Nonetheless, if an important scientific question is being asked and the experiment does not involve inflicting pain or suffering on the animal, experimentation with animals (at least nonprimates), especially genetic experimentation, is generally supported by both the public and scientific communities. The disgraced Korean cloning researcher, the veterinarian Woo Suk Hwang, who fabricated the results of his human cell cloning experiments, announced in 2014 that he had resumed work on dogs and other animals, and has cloned hundreds of dogs. In one experiment he claims to have cloned a dog that has symptoms of Alzheimer disease to see if he can develop an animal model to study this devastating human disorder. We are skeptical of his work, but cloning can produce genetically identical animals that are useful in genetic research. Nonetheless, as Ian Wilmut has stressed, "Cloning does not confer exact replication. The dialogue between the genes and their surroundings . . . controls the development of an organism from a single cell into a sheep—or indeed into a human being or an oak tree."

The ethics of research on humans, including genetic research, have been articulated primarily in reaction to horrific scandals. One of the worst involved the murder and torture of twins by the Nazi physician Josef Mengele at Auschwitz-Birkenau concentration camps during World War II. Mengele specialized in genetics, working under the supervision of Germany's leading geneticist, Otmar von Verschuer, and his genetics research at Auschwitz was funded by the German Research Council. Both men had a particular interest in doing genetic studies on twins. Survivors of Auschwitz recall Mengele shouting when the transports arrived, "*Zwillinge, zwillinge, zwillinge*" (twins, twins, twins). His biographers say that Mengele's main genetics goal at Auschwitz was to establish, using twins, which attributes (including blue eyes and blond hair) and disabilities are genetically inherited and which are determined by lifestyle and environment. Little, if anything, he did can be labeled science, however, and most of his "studies" were bizarre and unconscionable. Virtually all ended with the simultaneous murder and dissection of the twin pair.

The Nuremberg Code, which among other provisions requires the voluntary, competent, and informed consent of the potential research subject, was fashioned by U.S. judges at the Doctors' Trial at Nuremberg after the war, where some of the major physician war criminals were tried for murder and torture under the guise of human experimentation. Mengele, the most notorious of the murderous experimenters, escaped to South America after the war and was never apprehended. Needless to say, the Nazi twin experiments and Nazi belief in the racial (genetic) superiority of the Aryans did little to foster public faith in genetic research or to distinguish genetics from eugenics, the selection of desirable human traits to improve the human race, usually by selective breeding.

The Nazi twin murders and tortures in the guise of experiments had no justification, but legitimate twin studies had long been advocated in genetics research. Francis Galton first described the method of twin studies in 1875 as a way of disentangling the relative environmental and genetic influences on individual traits, behaviors, and disease states. Identical twins (termed monozygotic twins) result from fertilization of a single egg with a single sperm; fraternal twins (termed dizygotic twins) result from fertilization of two eggs by different sperm at the same time. In classical twin studies, comparisons are made for correlations of a disorder (for example, cancer or diabetes) or a quantitative characteristic (such as height or blood pressure) between identical twins and fraternal twins. If both members of a twin pair share a trait, they are *concordant;* if they don't share the trait, they are *discordant*. Because identical twins are (almost) genetically identical, differences between them should be due to environmental effects. Fraternal twins are also a convenient comparison; their environmental differences should be similar to those of identical twins, but their genetic differences are the same as those between siblings in that they share only 50 percent of their DNA.

Using twin studies to help clarify the "nature versus nurture" debate is nonetheless complicated because identical twins never share an identical environment. For example, during pregnancy, blood flow to each twin differs. How much a disease is influenced by genetics compared to environmental factors varies by disease. Perhaps the most impressive twin study to date involved data from 53,666 identical twins from the United States, Sweden, Finland, Denmark, and Norway. Using registry data that included twenty-four diseases, the researchers found that rather than one twin helping to predict the other's disease risk, the twin had about the same risk for these diseases as the general population. Commenting on the study, professor of genetics at Harvard David Altshuler observed that "even if you know everything about genetics, prediction will remain probabilistic and not deterministic." He thought this was true because disease is affected not only by genetics but also by behavior, the environment, and random events. One of those random events involves changes in DNA itself.

Until recently it was believed that identical twins had identical DNA throughout their lives. However, new genomic technologies have shown that there are often differences in the DNA of identical twins. These differences include both additional DNA segments (duplications) and lost segments of DNA (deletions) on certain chromosomes, a genetic state called copy number variants, or CNVs. Such variation is a natural occurrence that accumulates with age in everyone. Our genomes experience gains and losses of genetic material over time. Thus, as identical twins age, their somatic (body) cells diverge in CNVs. The extent to which CNVs contribute to human disease is unclear and remains an important area of ongoing research, as discussed in chapter 6.

Studies of identical twins are also shedding light on interactions between genome and environment at the molecular level. *Epigenetics* refers to a rapidly growing field that investigates heritable alterations in gene expression caused by mechanisms other than changes in the DNA sequence. Environmental factors can modify genes and their actions without changing the nucleotides (the *A*s, *T*s, *G*s, and *C*s) themselves, an issue we now explore in the context of pregnancy.

Genetics and Environment in Pregnancy

Altering the environment can significantly affect genomic expression as early as embryonic life. Sherman saw Erin and Sean during their first pregnancy because Erin had a routine blood test at about sixteen weeks of pregnancy that showed an abnormally elevated level of maternal serum alpha-fetoprotein, or MSAFP. Further testing determined that the fetus had spina bifida, a birth defect in which the bones of the spine (vertebrae) and overlying tissues do not form properly around part of the spinal cord. Their fetus had a sack of fluid containing part of the spinal cord, called a myelomeningocele, bulging through an opening in the back. After detailed counseling about options, Erin and Sean decided to continue the pregnancy. Their child, Sean Jr., required surgery shortly after birth to close the defect. To prevent hydrocephalus (increased cerebrospinal fluid pressure within the brain), a shunt—or drainage tube—was placed in Sean Jr.'s head. At four years of age, Sean Jr. had normal intelligence, urinary incontinence, and required forearm crutches to walk.

Erin and Sean planned to have a second child and consulted Sherman about the risk of having another child with spina bifida and what measures could be taken to prevent it. They were told that the recurrence risk for spina bifida, anencephaly (absence of a major portion of the brain), and related malformations (collectively called neural tube defects, or NTDs) was 2–5 percent. Most NTDs are caused by multiple genes in combination with environmental factors. A key environmental influence in the development of NTDs appears to be diet, most importantly folic acid (a B vitamin) intake. It is recommended that all women who might become pregnant take 400 micrograms of folic acid daily, starting before conception and continuing through the first trimester. For women such as Erin, who have had a previous pregnancy with an NTD, a tenfold higher dose of folic acid supplementation is recommended. This could reduce the risk of having a child with an NTD by about 80 percent.

Three approaches to preventing NTDs have been recommended by the U.S. Public Health Service: improve dietary intake of folate-rich foods (for example, lentils, spinach, or black beans); use dietary supplements containing folic acid (400 mcg per day, unless there is a history of a prior NTD, in which case 4 mg per day is recommended); and fortify foods with folic acid. NTDs have been reduced by an estimated 26 percent in the United States by fortification of foods, but failure to eliminate NTDs suggests that factors other than maternal deficiency in folic acid are involved.

Genetics, Environment, and Diabetes

When Natalie married Craig, she was twenty-two years old, weighed 136 pounds, and was in good general health. Natalie is now forty-three, weighs 247 pounds, and has just been diagnosed as having type 2 diabetes and high blood pressure. Her sixty-four-year-old mother also has type 2 diabetes, weighs 238 pounds, and recently had a stroke that has affected her ability to walk and slurred her speech. Natalie's doctor told her that she has a "metabolic syndrome," a designation given to someone who has risk factors that occur together and increase the likelihood of coronary artery disease, stroke, and type 2 diabetes. These risk factors include increased blood pressure, high blood sugar levels, excess fat around the waist, and abnormal blood fat levels.

Four months ago Craig was laid off from his job as a road repair worker. The family's income is now dependent upon Natalie's job as a paralegal professional at a small law firm and Craig's unemployment insurance. Between her job and having to take care of her mother, Craig, and their three children, Natalie does not have time for exercise, and eating a healthier diet is more costly and less convenient than meals at fast food restaurants. She wonders whether her obesity and diabetes are simply her destiny, given her family history. She is concerned about her future health but also the health of her children and their risk of becoming obese and developing diabetes.

Diabetes (*diabetes mellitus*) is a disease in which there are high levels of a sugar (glucose) in the blood. When food is digested, it is broken down and glucose enters the bloodstream. Insulin, a hormone released from the pancreas, moves glucose from the bloodstream into fat, muscle, and liver, where it is used as a source of fuel. Once inside the cells, glucose is either used immediately for energy or converted to fat or glycogen (a long-term storage form of glucose). People with diabetes have high levels of blood glucose because their pancreas does not make enough insulin, their cells do not respond to insulin normally, or both. In the United States, diabetes is the leading cause of kidney failure, nontraumatic lower-limb amputations, and new cases of blindness among adults; it is a major cause of heart disease and stroke; and it is the seventh leading cause of death. Of course, our environment plays a major role. High-fat diets typical of fast food restaurants—what sociologist George Ritzer called the "McDonaldization" of society—have become the norm in Western nations and are spreading globally. In the United States, more than a third of all adults (36 percent) are obese; about 17 percent (12.5 million) of children and adolescents ages two to nineteen years are obese.

There are three major forms of diabetes: type 1, type 2, and gestational diabetes. *Type 1 diabetes* (formerly called juvenile-onset diabetes) can occur at any age but most commonly develops in children, teenagers, and young adults, and accounts for about 10 percent of diabetics. It results from the permanent destruction of most of the insulin-producing cells, called beta cells, in the pancreas (most often due to an autoimmune disorder in which the immune system mistakenly attacks and destroys the beta cells). *Type 2 diabetes* (formerly called adult-onset diabetes) makes up the majority of diabetes cases. Although the pancreas produces insulin, sometimes at higher levels than normal, the tissues of the body do not respond appropriately to insulin—called insulin resistance. Reduced physical activity, poor diet, and excess weight, particularly around the waist, are all associated risk factors. Among U.S. residents sixty-five and older, almost 11 million individuals, or about 30 percent, have this disease. *Gestational diabetes* is defined as high blood sugar that begins or is first recognized during pregnancy. It affects 2–10 percent of all pregnant women.

Natalie has type 2 diabetes, obesity, and associated health problems. Considering how common and costly diabetes is, it remains remarkably poorly understood. Can new knowledge about type 2 diabetes at the genomic level help us better understand the underlying causes of this disease and lead to "actionable" therapeutic and preventive strategies?

There are some rare single-gene disorders associated with diabetes for which improved diagnosis and treatments have been developed. Maturity-onset diabetes of the young (MODY) accounts for about 2 percent of all cases of diabetes and is caused by mutation in one of several genes. Mutations in a gene called *HNF1A* are the most common cause. Testing for *HNFA1* mutations is important not only for correct diagnosis but also to identify other family members who are mutation carriers. In the vast majority of individuals with type 2 diabetes, however, there is no mutation in a single gene responsible for the disease. Researchers have undertaken *genomewide association studies* (GWAS) to understand what genes affect diabetes. GWAS compare common genetic DNA variants among large numbers of individuals who have a condition (for example, diabetes) and those who do not have the condition to determine whether an association exists between specific variants and the condition. Although a valuable research tool, GWAS have not had much practical impact on clinical management. This is because the genetic variants that have been detected are associated with the disease, but they do not cause it.

From conception and throughout fetal development, environmental influences can affect and even permanently modify genes that predispose people to adult diseases. This applies not only to type 2 diabetes but to other conditions, including obesity, coronary artery disease, and even conditions such as osteoporosis, cancer, and psychiatric illnesses. More than twenty years ago, a group of epidemiologists in Southampton, England, led by David Barker observed that the smaller a baby was at birth the higher the likelihood that he or she would die of coronary artery disease as an adult. This was followed by their findings that malnutrition of the fetus during pregnancy leads to the development of cardiovascular disease as an adult. These observations were popularized as the "Barker Hypothesis," or fetal origin of adult disease. Later it was shown that low birth weight was associated with development of type 2 diabetes as well. How does this happen?

Even relatively mild changes in the intrauterine environment (for instance, diet, inflammation, toxins, and infections) can cause the fetus to be "programmed" to produce different physical or biochemical characteristics that can persist throughout a person's life span. One tragic example occurred during the Dutch famine in World War II and is dramatically described by Amsterdam researchers:

> To support the Allied offensive, the Dutch government in exile called for a strike of the Dutch railways. As a reprisal, the Germans banned all food transport [causing a famine in The Netherlands]. . . . At the height of the famine from December 1944 to April 1945, the official daily rations varied between 400 and 800 calories. . . .
>
> Throughout the winter of 1944–1945 the population had to live without light, without gas, without heat, laundries ceased operating, soap for personal use was unobtainable, and adequate clothing and shoes were lacking in most families. In hospitals, there was serious overcrowding as well as lack of medicines. *Above all, hunger dominated all misery* [emphasis added].

In follow-up studies, infants whose mothers had severe caloric restrictions in mid or late pregnancy were underweight, while those infants whose mothers endured the famine in early pregnancy had normal birth weights. Adults whose mothers were exposed to undernutrition during pregnancy developed health problems in later life, particularly reduced glucose tolerance indicating type 2 diabetes and blood lipid profiles characteristic of coronary heart disease. The effects of the undernutrition thus depended on its timing during pregnancy. The investigators concluded that maternal malnutrition during pregnancy affects the health of the child in later life. This fetal programming evolved as an adaptive response; in times of food shortage, metabolic adaptations that increase energy storage may be beneficial.

Clinical studies have shown that metabolic imprinting (the epigenetic programming of metabolism during prenatal life) caused by obesity and a diabetic intrauterine environment can be transmitted across generations. This helps explain the increase in obesity, gestational diabetes, and type 2 diabetes seen over the past several decades. For example, the children of Pima Indian women with diabetes have larger infants at birth and at five years of age. Maternal diabetes was also the strongest single risk factor for type 2 diabetes among Pima Indian youth, accounting for 40 percent of diabetes in that population. There is a direct relationship between maternal weight prior to pregnancy, weight gain during pregnancy, and large babies who develop type 2 diabetes in later life. This is a vicious cycle in which "diabetes begets diabetes": heavier mothers give birth to heavier daughters, who are at increased risk to be obese and develop type 2 diabetes during their reproductive years.

The dangers of fetal exposure have been the subject of increasingly hysterical media attention, which you should not take too seriously. An article in the *Chicago Tribune,* "Your Health Partly Wired in the Womb," reported that "possible threats to the fetus" include urban air pollution, which might cause alterations in chromosome structure; bisphenol A (a chemical present in many plastic bottles), which could change the way the fetus responds to estrogens, leading to subsequent advanced puberty; and even a mild case of the flu in the pregnant woman, which could predispose male offspring to heart disease after age sixty. Fetal exposure to the maternal stress hormone cortisol has been implicated in causing an epigenetic effect that "slows growth in most organs." In other words, if a pregnant woman constantly worries about the seemingly endless number of environmental exposures that could possibly harm her fetus, the worry itself could injure her fetus. Of course, pregnant women can't live in a bubble throughout pregnancy, so common sense should prevail: avoid recognized substances, such as alcohol, tobacco, and drugs, that can put your fetus at increased risk for birth defects, eat nutritious foods, exercise, reduce stress, and practice moderation.

What about Natalie? Unfortunately, genomic studies have thus far added little that can be used in clinical management to our understanding of cases of type 2 diabetes and obesity like Natalie's. A person's individual risk for type 2 diabetes or obesity reflects a "barcode" combination of susceptibility and protective genetic variants, which are influenced to a greater or lesser extent by "relevant" environmental exposures, primarily diet and exercise. But this is not all. In addition to complex gene-gene and gene-environment interactions, our health is also directly affected by the trillions of microbes that live in our bodies and make up our microbiome.

Your Microbiome

Your body is made up of about 10 trillion cells and is the home of 100 trillion microbes that live both within and on the surface of our bodies, including thousands of species of bacteria, yeasts, parasites, viruses, and others. *Microbiome* is the term for the totality of microbes, including their genomes, their gene products (such as proteins), and their unique interactions in a particular environment, or habitat, such as the gut, mouth, skin, or vagina. Human health and diseases are greatly influenced by the diversity of our microbiome, especially bacteria. We can't live without them.

The human microbiome contains at least 20 million unique genes (humans have about 22,000 protein-coding genes). From a cellular point of view, humans are more bacteria than they are human—bacteria outnumber human cells ten to one. However, because microorganisms are small, they make up only about 1–3 percent of the body's mass; a two-hundred-pound adult has two to six pounds of bacteria. Many of these organisms have not been cultured, identified, or otherwise characterized because their growth is dependent upon a microenvironment that has not been reproduced in the laboratory. However, new DNA-sequencing technologies and sophisticated computer bioinformatics now allow these microbes to be studied in great detail because we can distinguish between the DNA of humans and the DNA of microbes, so that only the bacterial genome is analyzed.

Bacteria that are absent are sometimes as important as the bacteria that are present. For example, *Streptococcus mutans* (*S. mutans*) is known to cause tooth decay by converting sugar to acid. Therefore, tooth decay can be considered an infectious disease. A synthetic antimicrobial, C16G2, has been shown to have robust killing efficacy against *S. mutans,* and even a single application of a mouth rinse containing C16G2 has been associated with reduced plaque, lactic acid production, enamel demineralization, and tooth decay.

Another example of the absence of a bacterium that caused disease is a sixty-one-year-old woman whom we will call Alice. She was referred to Alexander Khortus, a gastroenterologist at the University of Minnesota. Alice had suffered eight months of crippling diarrhea, which started shortly after she had been treated with antibiotics following back surgery that was complicated by pneumonia. Her diarrhea had become so severe that she had to wear diapers all the time. She had lost almost sixty pounds and needed to use a wheelchair. After several hospitalizations and numerous courses of antibiotics for a life-threatening bowel infection involving an organism called *Clostridium difficile* (*C. difficile*), treatment options were running out.

Khoruts decided that Alice needed a novel new treatment—a fecal transplant (transferring fecal matter from one patient to another). The donor was to be Alice's husband of forty-four years. Using colonoscopy, a stool sample from her husband that had been put through a blender and diluted was injected into Alice's colon (fecal transplants are now done by processing the fecal material and putting it into tablet form for consumption). In addition, colonoscopies were performed a week prior to the transplant and two weeks afterward. Khoruts and his associates were able to show that the disease-causing *C. difficile* in Alice's colon had been replaced by the normal gut bacteria *Bacteroides* from her husband. On the second day after the procedure, Alice had her first solid bowel movement. At her six-month follow-up, she reported once-daily formed stools. In a subsequent randomized trial of patients with recurrent infections and multiple episodes of diarrhea, the combination of antibiotics and fecal transplant was so successful (cures were achieved in fifteen of sixteen patients within two treatments) that the study was stopped early. An editorial accompanying the study concluded that "FMT [fecal microbiota transplantation] is now in the mainstream of modern, evidence-based medical practice." Science writer Ed Young has described the procedure as "spectacularly successful, far more than conventional antibiotics." Young also noted that the great advantage to finding health problems rooted in the microbiome is that unlike the human genome, we may be able to alter our microbiomes "through probiotics, fecal transplants or other means."

The gastrointestinal tract harbors a complex assemblage of microbial organisms that are essential for food digestion, absorption of nutrients, and development and regulation of the immune system. Bacteria stimulate the lymph tissues in the intestines to produce antibodies against pathogens. The immune system recognizes and fights harmful bacteria but does not react to helpful species of bacteria or a person's own tissues (if the immune system does attack the person's own tissues, it can lead to autoimmune diseases such as rheumatoid arthritis, lupus, inflammatory bowel disease, and type 1 diabetes). This tolerance develops in infancy and is called the "old friends" hypothesis.

In normal pregnancy, the fetus develops in the sterile environment of the uterus. At birth and rapidly thereafter, the infant is colonized with bacteria from its surrounding environment. When the baby is born through the vaginal canal, its first contact is with bacteria from the mother's vagina and anus, predominantly *Lactobacillus*. *Lactobacillus* is usually found not in the vagina but rather in the gut, where it produces enzymes that digest milk, but as pregnancy progresses toward term, *Lactobacillus* becomes prominent in the vagina. During a normal term delivery, the infant is covered with *Lactobacillus* as it goes through the vaginal canal, thereby preparing the infant to digest breast milk.

Approximately one in three babies in the United States are born by cesarean delivery. Cesarean babies are first exposed to and colonized by bacteria found on the mother's skin, predominantly *Staphylococcus*. These differences in exposure can be significant. For example, most cases of virulent and often fatal methicillin-resistant *Staphylococcus aureus* (MRSA) infections in infants follow cesarean deliveries. There is also accumulating evidence that these differences in intestinal bacteria play an important role in the development of the infant's immune system. Microbes in the intestines may be involved in the so-called "hygiene hypothesis," wherein lack of exposure to a variety of microorganisms during early life may result in the development of diseases such as asthma, type 1 diabetes, celiac disease, and food allergies, which appear more often in children who were born via cesarean delivery compared to vaginal delivery.

A more dramatic example of how things go seriously awry is when the intestines fail to discriminate between friendly and hostile bacteria; inflammation sets in that can progress to a condition called necrotizing enterocolitis (NEC), or death of the intestines. NEC is among the most common and devastating diseases in newborns. The estimated death rate from this condition is 20 percent. The excessive inflammation initiated in the highly immunoreactive intestine in NEC extends to other organs, including the brain, kidneys, lungs, liver, and heart.

Almost all infants born prematurely are treated with broad-spectrum antibiotics, which decrease the diversity of the intestinal bacteria and increase "less desirable" bacteria, thus facilitating the development of NEC. Recent studies suggest that giving these infants *probiotics* (live microorganisms, such as lactic acid bacteria and certain yeasts that confer health benefits) could help prevent NEC. Probiotics should be distinguished from *prebiotics*, which are nondigestible food ingredients, such as soluble fibers and oligosaccharides found in breast milk. Like probiotics, prebiotics have a beneficial effect on intestinal bacterial microbes and have been shown to help prevent NEC in premature infants.

In developed countries, the average child has received ten to twenty courses of antibiotics by the time the child is eighteen years old. Antibiotic treatment can prevent or cure serious diseases and may be lifesaving, but overuse of antibiotics carries substantial risks to our health. Sometimes our friendly bacterial flora never fully recovers after treatment with antibiotics. Some of these friendly bacteria provide important benefits. For example, the *Bacteroides* species that live in the colon are needed to synthesize vitamin K, which is required for blood coagulation and also helps resist invading organisms.

The role of microorganisms is not always straightforward. *Helicobacter pylori* (*H. pylori*), a corkscrew-shaped bacterium, for example, is found in half the stomachs in the world. This is astounding because gastric (stomach) fluid is composed mostly of hydrochloric acid with acidity of a pH between 1.5 and 3.5. *H. pylori* is associated with increased risk of gastritis (inflammation of the lining of the stomach), peptic ulcers, and stomach cancer. However, children without this bacterium are more likely to develop asthma, hay fever, and skin allergies. Eradication of *H. pylori* from people's stomachs has also been linked with an increase in gastroesophageal reflux disease (GERD) and esophageal cancer.

The story of how *H. pylori* was discovered as the major cause of peptic ulcers seems like the plot of a mad scientist movie. Until the 1980s medical dogma held that gastritis, stomach ulcers, and duodenal ulcers were caused by overproduction of gastric acid resulting from chronic stress. As one article put it in 1967, "The mothers of ulcer patients tended to have psychogenic symptoms, and to be striving, obsessional, and dominant in the home; fathers tended to be steady, unassertive, and passive." The dictum was "no acid, no ulcers," which led to treatments to neutralize acid, such as bland diets and antacids. In 1981, Barry Marshall, an Australian internist, began working with a pathologist, Robin Warren, at the Royal Perth Hospital. Taking biopsies from ulcer patients, they discovered that the gut was overrun by a newly identified bacterium, *Campylorbacter pyloridis,* later called *H. pylori*. This was an astonishing observation because it meant that ulcers were an infectious disease that could be treated with antibiotics.

The finding was dismissed by mainstream gastroenterologists. In Marshall's words, "I knew that they were mostly making their living doing endoscopies on ulcer patients. So I'm going to show you guys. A few years from now you'll be saying, 'Hey! Where did all those endoscopies go?'" The breakthrough came in 1982. Marshall and a volunteer drank a broth containing the suspected bacteria, then underwent endoscopies showing that they had both developed gastritis and had biopsies from which *H. pylori* was reisolated. The connection between *H. pylori* and ulcers was subsequently confirmed by large epidemiological studies. For their discovery, Marshall and Warren were awarded the Nobel Prize for Medicine in 2005. In an interview ten years later, Marshall opined that other serious diseases may turn out to have an infectious cause, saying, "As far as I am concerned, everything is environmental until you convince me that it is genetic." This may seem extreme, but it underlines the power of environment, including our microbiome, to affect our health, and that is why we opened the chapter with it.

The most common clinical conditions currently studied in the context of the microbiome are obesity and type 2 diabetes. As we have discussed, obesity and type 2 diabetes are caused by a combination of genetic susceptibility and environmental and lifestyle factors. Compared to lean individuals, the bacterial composition in the intestines of obese individuals extracts more energy (calories) from their diet, increases storage of fat, and leads to insulin resistance. For example, eradicating *H. pylori* with antibiotics disrupts the stomach's steady production of a hormone called ghrelin, which is released into the bloodstream and in essence tells the brain to keep eating. Farmers have long known that continuous subtherapeutic doses of antibiotics cause animals to gain weight with less feed. Returning the gut to a healthy bacterial state, possibly with fecal transplants or the use of pre- or probiotics, has been suggested as a pathway to reduce obesity and improve insulin sensitivity in type 2 diabetes.

Another area of the body where there is an unexpected diversity of bacteria is on the skin, our first line of defense against illness and injury. The most diversity is seen on the forearm and the least diversity behind the ear. The bacterial composition of our underarms is different from the bacteria that live between our toes, which explains why they smell different. Bacteria on our skin have many beneficial effects, including converting oils into moisturizers that keep our skin supple and prevent cracking. Of course, we know that not all bacteria on our skin are good for us; some bacteria cause infection, hence hand washing is important.

In a *New Yorker* article, Michael Specter recounts how Andrew Goldberg, an ear, nose and throat specialist at the University of California, San Francisco, cared for a patient (whom we'll call Colin) with a chronic infection in his left ear. Doctors had tried numerous treatments, including several types of antibiotics and antifungal drops, without success. One day Colin walked into the clinic with a smile on his face, saying that he hadn't felt so good in years. Goldberg examined his ear and confirmed that it looked great. Colin said, "Do you want to know what I did? . . . I took some wax from my good ear and put it into my bad ear, and in a few days I was fine." Years later Goldberg came to understand that normally earwax contains many bacterial species, and that antibiotics might have adversely changed the bacterial composition in Colin's bad ear. By reestablishing the normal ear bacteria by an "earwax transplant," Colin cured himself.

For the past century, doctors have waged "war" against bacteria with antibiotics. But our view of microbes is changing rapidly. As Julie Segre, a senior investigator at the National Human Genome Research Institute, put it, "I would like to lose the language of warfare. . . . It does a disservice to all bacteria that have co-evolved with us and are maintaining the health of our bodies." The new health model is "medical ecology," and at least some physicians are beginning to see themselves as "microbial wildlife managers."

In this chapter we have seen that, as Dolly's creator, Ian Wilmut, put it more than a decade ago, there is more to life than genes. "Genes operate in constant dialogue with their surroundings" a dialogue that begins at conception and continues after birth and throughout life. Today, he would have to add our microbiome to the mix, as we are learning more and more about our microbiome's importance to our daily health. The balance and interaction of our genes and the environment determine if and how genes will be expressed. We are dealing not with a "nature versus nurture" world but with a "nature and nurture" world. This interaction affects almost every aspect of our lives, but perhaps most importantly to our physicians, it affects the way we are likely to react to specific drugs. In the next chapter, we turn to what is widely considered the most promising clinical application of the evolving field of genomics: pharmacogenomics.

WHEN THINKING ABOUT NATURE VERSUS NURTURE, CONSIDER THESE THOUGHTS

It is not nature versus nurture, but nature and nurture (and our microbiome).

Our environment plays a powerful role in determining our health and the health of our children and our grandchildren.

A fast-growing area of genomic studies involves the human microbiome and its effect on our health.

CHAPTER 4

Pharmacogenomics

We must not see any person as an abstraction. Instead, we must see in every person a universe with its own secrets, with its own treasures, with its own sources of anguish, and with some measure of triumph.

—*Elie Wiesel,* preface to *The Nazi Doctors and the Nuremberg Code* (1993)

Rebecca Skloot opens her immensely popular book, *The Immortal Life of Henrietta Lacks,* with this statement from Elie Wiesel reflecting on the uniqueness of every human being. Genomics, of course, underlines that uniqueness. Skloot's book is about Henrietta Lacks, the woman whose cervical cancer tumor was the source of the famous HeLa cells, an immortal cell line that has been the backbone of laboratory research around the world for more than sixty years (figure 4.1). Portions of her tumor were removed for study at the segregated Johns Hopkins Hospital in 1951. Her story is a compelling portrait of an extended black family and their race- and poverty-based interactions with the white medical profession from the 1940s to the present. Genetics is never far from the surface. The physician who is sometimes called the "father of American genetics," Victor McKusick, for example, began his own research on the Lacks family in 1973. McKusick wanted blood samples from the surviving Lacks

4.1 Henrietta Lacks. With permission of Rebecca Skloot, *The Immortal Life of Henrietta Lacks* (New York: Crown Publisher, 2010).

family members to try to locate genetic markers that could be used to identify HeLa cells in various laboratory experiments. No informed consent for the blood drawing was obtained from the family members, who believed their blood was being tested to see if they had the same cancer Henrietta had died of more than twenty years previously. Lacks was thirty-one years old when she died. No consent had been obtained to use her cancerous cells in 1951 either.

For her book, Rebecca Skloot interviewed Susan Hsu, the member of McKusick's team who drew the blood samples from the Lackses. Hsu said she felt very bad that the Lacks family didn't understand why she was collecting their blood more than twenty-five years ago. She added that she could learn much more today from their blood, because of advances in DNA technology, than she could in 1970. She remarkably asked Skloot if she would be willing to ask them if they'd give blood again for her research: "If they are willing, I wouldn't mind to go back and get some more blood."

Lacks's daughter told Skloot that she thought it unfair that even though her mother's cells helped create the extremely lucrative biotechnology industry, her family lacked health insurance and struggled to pay their medical bills. Law professor and expert on racial discrimination Dorothy Roberts describes the Lacks story as a reflection of the "horrible history of medical exploitation and neglect of African Americans" on the basis that they are genetically inferior to whites. But, as she properly underlines, the Henrietta Lacks story also refutes that belief: "Her cells, although they came from a black woman, helped to improve the health of human beings the world over and testify to our common humanity."

The tendency to treat differences in socially constructed racial groupings as genetics based has haunted genetic medicine since its beginnings, and we continue to be plagued by racism in genomics. In this chapter, we attempt to place pharmacogenomics—the use of genomics to determine which drug at which dose to give an individual patient—in the context of racism and racial disparities in American medicine. We then compare the way foods and drugs are affected by our genomes and discuss the usefulness of genomic screening to determine what drugs may be helpful for your condition.

Pharmacogenomics can usefully be described as selecting a drug and its dose on the basis of an individual patient's genomic profile. The goal is to maximize the drug's efficacy and minimize its toxicity. It is not, however, "personal" in the sense that we are trying to select a specific drug and dose just for you. Instead, pharmacogenomics is about selecting a drug and dose for the group of patients with whom you share specific genetic characteristics. These genetic similarities can be used to categorize or stratify patients in groups that are likely to react the same way to specific drugs. Thus *stratified medicine* is a more accurate term than *personalized medicine* for genomic medicine. Identifying the genes that control drug effectiveness also permits clinicians to improve treatment by using genomics rather than physical characteristics, such as skin color. This is important because of the history of racism in American medicine, complete with unscientific views of the differences among people of different colors.

Genetics and Racism

Racism continues to plague pharmacogenomics at the clinical level. For example, a 2014 study concluding that black women with breast cancer were 40 percent more likely to die of the disease than white women showed that survival differences were not because of any genetic difference, but resulted from "systematic racism," which delayed diagnosis and treatment. This is not just true for breast cancer. Racial disparities in health care and health outcomes in almost all areas of medicine are a function not of biological differences between minority and majority populations, but rather differing environmental factors, including racism, poverty, lack of health insurance, and lack of access to decent medical care. Nonetheless, because genetics has been widely misused to suggest a biological basis for racial differences, it is critical that doctors and patients alike understand the relationship between genes and skin color. The stories of the Tuskegee Syphilis Study, like the story of Henrietta Lacks, are emblematic of past medical practices that continue to incite distrust on the part of minorities in medicine, distrust that threatens to undermine even the promise of pharmacogenomics.

"The United States government did something that was wrong—deeply, profoundly, morally wrong. It was an outrage to our commitment to integrity and equality for all our citizens." President Bill Clinton spoke these words at a White House ceremony—this one held three years before a more celebratory White House ceremony announcing the completion of the draft of the human genome. The occasion, in 1997, was a formal apology to the surviving victims of what is known as the Tuskegee Syphilis Study. The study was based on racism: a belief that black men were so different biologically from white men that syphilis in blacks was a different disease than syphilis in whites. The study followed the progression of syphilis in a cohort of black males until they died. The subjects were lied to about the study and neither informed of their disease nor treated for it. Medical historian James Jones described the study as the "longest nontherapeutic experiment on human beings in medical history." It ran from 1932 to

1972. The Tuskegee Syphilis Study remains a hideous blot on American medicine. In words that remind us of Elie Wiesel's commentary on the Nuremberg Doctors' Trial, quoted to open this chapter, survivor Herman Shaw reflected, "We were treated unfairly and to some extent like guinea pigs. We were not pigs. We were not dancing boys as we were projected in the movie, *Miss Evers Boys*. We were all hard working men, not boys, and citizens of the United States. The wounds that were inflicted upon us cannot be undone."

There are no genes that code for race, but there are genes that appear with greater frequency in different ethnic groups. It is disturbing that race and ethnicity are often used interchangeably in the scientific literature. Race, a social construct, not a scientific or genetic one, and is a poor proxy for genetics. Even for genes that appear frequently in specific ethnic groups, it is far more scientific and accurate to directly identify a gene or genes that affect drug metabolism. A leading geneticist, Arno Motulsky, is credited with opening the race/gene/drug interaction exploration in a 1957 article on "genetically conditioned drug reactions." He concluded (using race and ethnicity interchangeably) that "since a given gene may be more frequent in certain ethnic groups, any drug reaction that is more frequently observed in a given racial group, when other environmental variables are equal, will usually have a genetic basis." The Tuskegee study was still going on, so it is probably not surprising that the major study Motulsky cited for the proposition that blacks responded differently than whites to the malaria drug primaquine was one done exclusively on black prisoners in the early 1950s. With so little known about genetics at the time, Motulsky thought that observing drug reactions in humans could "contribute to the progress of human genetics in general." Science is now rightly in the process of trying to reverse this order, using human genetics to predict drug reactions.

Pharmacogenomics got a big boost from the 2000 White House announcement of the draft human genome. A number of the world's most prominent geneticists announced that the project would once and for all put an end to assertions that human groupings are fundamentally different at the genetic level. Chris Stinger of London's Natural History Museum observed, "We are all Africans under the skin." Other geneticists said, "Race is only skin deep" and "There is nothing scientific about race: no genes of any sort pattern along racial lines." Craig Venter summed it up: "Race is a social concept, not a scientific one. We all evolved in the last 100,000 years from the same small number of tribes that migrated out of Africa and colonized the world." Geography matters, because people who live in the same area are more likely to have children together and perpetuate common genes. Race does not matter, because two black people are no more likely than a white person and a black person to share the same genes. The most scientifically valid approach is therefore to directly identify the relevant genes or variants.

In 2001, at the World Conference on Racism in South Africa, George suggested that it would be cold comfort to simply replace racism (the belief that fundamental human characteristics are determined by race) with "genism" (the belief that fundamental human characteristics are determined by genes). We both continue to believe that this is a danger, but neither of us was ready for the first round of pharmacogenomics, which was dominated by using race rather than genetics as a rationale for developing the drug BiDil (discussed later). As Eduardo Bonilla-Silva, author of *Racism Without Racists,* observed, "Race, like Freddy Krueger, keeps coming back after we believe we killed it." In a climate in which racial disparities continue to plague contemporary medicine and black patients continue to be suspicious of the white medical establishment, we should not be shocked.

The death of race-based genetic medicine has been hard for some to accept. In 2000, for example, a small Massachusetts biotech firm, NitroMed, announced that it had approval from the FDA to conduct a clinical trial with its new drug, BiDil. What made BiDil controversial was that it was a heart failure medication designed exclusively for use by African Americans. Following a clinical trial that involved only blacks (how is this possible in the twenty-first century?), the FDA approved BiDil "as the first drug with a race-specific indication: to treat heart failure in a 'black' patient." BiDil was not a new biotech drug but simply two ("bi") generic blood vessel dilators ("dil") combined into one pill. Nonetheless, its approval was hailed as a major milestone in pharmacogenomics. As *Science* put it, "By backing BiDil, the FDA panel gave another push to pharmacogenomics, an approach that promises to revolutionize both drug discovery and patient care."

Jonathan Kahn, the historian of BiDil (and author of *Race in a Bottle,* from which much of our material on the drug is taken), has emphasized that BiDil is in no way a genomic or pharmacogenomic drug—as no related genes were even looked for, let alone identified. Instead, as the chair of the FDA advisory panel, Steven Nissen, put it (without irony), "We're using self-identified race as a surrogate for genetic markers." He continued, "We are moving forward to genome-based medicine. It's going to happen." Kahn notes that the lack of any genetic basis for the drug actually helped to market it: instead of having to screen patients for a gene or genes, it could simply be marketed to African Americans. In his words, "Medical researchers might see race as a surrogate to get at biology in drug development, but corporations use biology as a surrogate to get at race in drug marketing." Although ultimately a failure in the medical marketplace, as the most discussed race-based drug in the world, BiDil teaches us that racial stereotyping is not dead, even in medicine and science, and that genomics has sufficient public appeal that it can serve as a cover story to promote even nongenomic-related pharmaceuticals to both the FDA and the public. This is fake genomics.

Another attempt to inject race into genomics was *New York Times* reporter Nicholas Wade's *A Troublesome Inheritance: Genes, Race and Human History.* He argues, for example, that culture is at least partly genetic, so genetics can help explain why, for example, "American institutions do not transplant so easily to tribal societies like Iraq or Afghanistan." This implausible argument is in the same category as the argument that genes (rather than access to medical care, poverty, and other factors) can explain why black women have worse outcomes from breast cancer treatment than white women. As reviewer H. Allen Orr put it, "What about all those other differences [i.e., other than genes] in history, language, distribution of wealth, religion, educational attainment, ravages of war, arable land, resentment toward perceived invaders, and so on? Among these factors, I suspect that genes are perhaps the one most *similar* between American and Afghan societies." Bad or unscientific genomics is relatively easy to spot and debunk. But what does good genomics look like in the prescription drug field? The story of warfarin answers this question.

Pharmacogenomics and Warfarin

On NBC's *Today Show*, Serena Williams, one of the all-time greatest professional tennis players, described how on February 18, 2011, she was rushed to a hospital suffering from swelling of her leg and severe shortness of breath. (Serena had cut her foot on a piece of glass the year before, a cut that had required two surgeries and her being in a leg cast for ten weeks.) The doctors did a CAT scan that showed Serena had experienced pulmonary embolisms—blood clots in the lungs—a potentially life-threatening condition. The pulmonary embolisms were caused by blood clots in her leg breaking loose and traveling to her lungs. She was discharged on anticoagulant drug injections and was able to attend parties at the Academy Awards on February 27. The

4.2 Serena Williams hitting a backhand shot during the 2012 Wimbledon Championships on Day 10 in the Semifinals versus Victoria Azarenka. *Wikimedia Images,* July 5, 2012.

next day she was readmitted to the hospital because she had developed a grapefruit-sized collection of blood underneath her skin (a hematoma) that required surgical drainage. Later, Serena was switched to an oral anticoagulant drug, warfarin, and had to wear a boot for an additional ten weeks. Thanks in part to warfarin, she was able to resume her spectacular tennis career (figure 4.2).

Approximately half a million people have a pulmonary embolism in the United States annually. If untreated, 30 percent of them would die. Drugs that diminish the ability of blood to clot, called anticoagulants, are a mainstay of treatment for pulmonary embolism. Like Serena, patients are usually given an injectable anticoagulant as an initial treatment, then converted to pills. The most commonly prescribed anticoagulant pill is warfarin. Then secretary of state Hillary Clinton was prescribed warfarin when she was hospitalized at the end of 2012. She had a blood clot just behind her left ear (a right transverse sinus thrombosis) as a result of hitting her head after she fainted and fell a few weeks earlier. Her physicians announced, "She will be released once the medication dose [of warfarin] has been established."

Warfarin works by inhibiting an enzyme that uses vitamin K, the vitamin which is necessary for the production of several blood-clotting proteins in the liver. It was initially marketed as a rat poison and is still used for this purpose. Saying that warfarin is a "blood thinner" is a misnomer, since it does not affect the thickness (viscosity) of the blood; rather it reduces the degree of clotting. Warfarin is not only used to prevent and treat embolisms. It is also used for people with certain types of irregular heartbeats, people with replacement or mechanical heart valves, and people who have suffered a heart attack. In each of these conditions, warfarin is used to help prevent a blood clot from developing and traveling to the brain to cause a stroke, or to the lung to cause respiratory failure.

Determining the optimal dose of warfarin is tricky because of marked and often unpredictable dosing variation among individuals. Giving too little warfarin does not protect against the development of clots, whereas giving too much can cause bleeding problems. The primary way a physician monitors the anticoagulant effect of warfarin is by trying to maintain the drug dose within a narrow therapeutic range.

The difficulty of maintaining the right dose of warfarin is evidenced by warfarin being one of the drugs most often responsible for emergency room visits, most commonly hemorrhage (too much warfarin) or blood clots and stroke (too little warfarin). It has been established that two genes primarily affect how individuals respond to warfarin.

It should be emphasized that even though genetic testing can provide physicians with relevant dosing information, there is no evidence that genetic testing was done for either Serena Williams or Hillary Clinton. The point is not that such testing is irrelevant; rather, it is time consuming, and its impact on treatment is currently marginal. With or without the genetic profile, physicians will still use trial and error to establish the proper dosage of warfarin for a particular patient. Use of genetic information to help inform the dosage decision will only routinely occur only after patients have their whole-genome profile included in their electronic health records. Access to relevant genetic information will then be easy and quick.

The FDA approved a labeling change for warfarin (Coumadin) in 2007, explaining that people's genetic makeup may influence how they respond to the drug and highlighting "the opportunity for healthcare providers to use genetic tests to improve their initial estimate of what is a reasonable warfarin dose for individual patients. Testing may help optimize the use of warfarin and lower the risk of bleeding complications of the drug." The FDA commissioner said, "Today's approved labeling change is one step in our commitment to personalized medicine. By using modern science to get *the right drug in the right dose for the right patient*, the FDA will further enhance the safety and effectiveness of the medicines Americans depend upon [emphasis added]." No requirement for genetic testing was included in the labeling change. Shortly thereafter, the FDA cleared the way for the first genetic tests for warfarin metabolism.

Medicare announced in 2009 that it would not pay for genetic tests for warfarin metabolism (which cost $50 to $300) because there was not enough evidence that use of the tests actually improved patients' health. The agency did say that it would pay for the tests as part of clinical trials to gather such evidence. The evidence has not yet been found. For example, a study published in 2014 randomly assigned one thousand patients to warfarin on the basis of clinical variables only or on the basis of clinical variables plus genotype information. Genotyping did not improve anticoagulation control during the first month of therapy. We think, nonetheless, that effectiveness will change when physicians become more comfortable using genetic information to inform dosing decisions.

The activity of other drugs has also been linked to specific genes. In the summer of 2012, for example, the FDA issued a warning when three children died and another child had life-threatening respiratory depression after taking routinely prescribed doses of codeine following surgery to remove tonsils. Health care professionals and parents were cautioned that when prescribing codeine-containing drugs, the lowest effective dose for the shortest time should be used. Codeine is used to treat mild to moderate pain but is also found in some cough suppressants. Once in the body, codeine is converted into morphine in the liver by an enzyme called CYP2D6. The *CYP2D6* gene that encodes this enzyme has at least eighty known variants. These variants are grouped to classify how well individual patients will metabolize codeine. At one extreme, "poor metabolizers" may not get the pain relief expected from codeine. At the other extreme are "ultrarapid metabolizers." After receiving normal doses of codeine, their livers quickly convert codeine into dangerously high blood levels of morphine that can lead to death. The three children who died after receiving codeine were likely ultrarapid metabolizers. In one tragic (and highly unusual) case, a mother who was subsequently found to be an ultrarapid metabolizer took codeine for postpartum episiotomy pain while she was breast-feeding. Her thirteen-day-old baby died of morphine "poisoning" with a serum concentration over thirty times higher than typically seen in babies whose mothers who are breast-feeding while receiving codeine.

Genetic testing can also help prevent organ rejection. This was true even in the early 1950s when transplanted kidneys were consistently rejected until identical twins were used as donor and recipient. Genetic tests to confirm that twins were identical did not exist at the time. Instead, genetic identity was determined much less scientifically on the basis of blood groups, fingerprints, and skin grafts between the twins.

About 18,000 Americans annually receive kidney transplants, one-third from living donors. For people with end-stage kidney disease, a kidney transplant offers enhanced quality of life. The main obstacle to more transplants is the shortage of cadaver organs, which would be helped if more Americans signed organ donor cards—something you should seriously consider.

Whichever protocol a transplant center uses, all kidney transplant recipients require lifelong immunosuppression to prevent rejection, just as many patients who have had blood clots require lifelong use of an anticoagulant. Perhaps the most commonly used drug for immune suppression is azathioprine (Imuran). An enzyme called thiopurine methyltranferase (TPMT), encoded by the *TPMT* gene, controls azathioprine activity. Approximately 10 percent of the population inherit one nonfunctional *TMPT* gene variant, conferring intermediate TMPT enzyme activity, and 0.3 percent inherit two nonfunctional variants for low or absent activity. Patients with intermediate TPMT activity may be at increased risk of bone marrow suppression—a decrease in cells responsible for providing immunity, carrying oxygen, and ensuring normal blood clotting—if receiving conventional doses of azathioprine. Patients with low or absent TPMT activity are at increased risk of developing severe, life-threatening bone marrow toxicity when receiving conventional doses. Analysis of *TPMT* gene variants can provide physicians with valuable guidance for either not prescribing azathioprine or adjusting its dose. In clinical practice, enzyme activity and genotyping are often used as complementary tests. Genotyping can tell us how we will likely react to specific drugs. It can also tell us how we might react to specific foods.

Food and Genes

Drugs and food are absorbed, metabolized, and excreted by bodily processes involving multiple genes, as well as environmental influences, including the types and amounts of food we eat, and the types and doses of drugs we take. The emerging field of nutrigenomics, which aims to identify the genetic factors that influence the body's response to diet and studies how bioactive ingredients of food affect gene expression, helps us understand the current state of pharmacogenomics. For centuries it has been recognized that tastes in food are not always inherited. Renaissance writer Michel de Montaigne, for instance, observed about his own appetite: "I am not excessively fond of either salads or fruits, except melons. My father hated all kinds of sauces; I love them all." He was also against regimented diets, writing, "Let us leave the daily diets to the almanac-makers and the doctors." He argued that the whole point of food was enjoyment, and was opposed to treating food like medicine. But foods can directly affect our health.

Some Mediterranean, African, and Asian people who ingest fava beans become seriously ill with high fever, weakness, jaundice, vomiting, and diarrhea. The condition is called favism. More than fifty years ago, it was determined that favism was caused by deficiency of an enzyme called glucose-6-phosphate dehydrogenase (G6PD), which is encoded by a gene on the X chromosome; hence the condition almost exclusively affects males. More than 140 mutations of the *G6PD* gene have been described that can result in episodes of severe acute hemolytic anemia (breakdown of red blood cells) and rarely in kidney failure and death.

In an episode of *MASH*, Max Klinger, a corporal of Lebanese descent, is trying to get a psychiatric discharge from the army. He is accused of faking exhaustion and back pain but is ultimately diagnosed as having G6PD deficiency resulting from the antimalarial drug primaquine. The black prisoners in the primaquine study (the study cited by Motulsky at the beginning of this chapter) who developed severe anemia at large doses also likely had G6PD deficiency. The study was based on the mistaken belief that black skin color, rather than ancestral geography, determined the medication's side effects. G6PD deficiency affects more than 400 million people worldwide. It is most common in areas where malaria is endemic, suggesting that the deficiency is protective against malaria. Knowledge that an individual has G6PD deficiency is important when considering the use of commonly prescribed drugs, high doses of vitamin C, and even aspirin.

Taste for certain foods, such as sushi or gefilte fish, may be acquired, but taste may also be genetic. Some of us really can't stand the taste or even the odor of particular foods, such as brussels sprouts, cilantro, or broccoli, no matter how well they are prepared or how hard we try. These and other food and odor aversions may be due, at least in part, to genetic factors. Differences among individuals in perception of the bitter taste of vegetables, as well as bitter beverages—coffee, grapefruit juice, quinine in tonic water, and alcohol—have been associated with variants in more than two dozen bitter receptor genes, collectively called *TAS2R*s. Some people cannot perceive table sugar (sucrose) in liquids, because of variants in the sweet receptor gene *TAS1R3*.

After eating asparagus, some of us smell a varying pungent sulfurous odor in our urine, like cooked cabbage. This effect is very rapid, taking as little as fifteen to thirty minutes. Two reasons for variations in this odor have been proposed. The first is that there are differences in the amount of odorant people produce. The second is that people vary in their ability to smell. Both reasons likely play a role. Differences in how people digest and metabolize asparagus and variations in the ability to smell the odor are due to differences in the DNA sequence at a single nucleotide in the gene *OR2M7,* which encodes a receptor for smell. Taste and smell are both complex. Genetic variations can only provide part of the answer, because of the contribution of environmental, behavioral, and cultural influences—which means we can't explain, for example, why Montaigne liked all the sauces his father hated.

What does all this have to do with pharmacogenomics? We have learned a lot about genes and nutrition but not enough to make health claims that genetic testing can give you enough information to develop a personalized diet. In early 2014, for example, the Federal Trade Commission (FTC) reached a consent agreement with GeneLink, in which the company agreed to stop marketing its "genetically customized nutritional supplements," which it claimed it could personalize to fit "each consumer's unique genetic profile" (based on an assessment of DNA obtained from a cheek swab provided by the consumer). The FTC announced under the proposed settlement that GeneLink was prohibited from claiming that any food would treat, prevent, or reduce "the risk of any diseases, including diabetes . . . by modulating the effect of genes, or based on a consumer's customized genetic assessment—

unless the claim is true and supported by at least two adequate and well-controlled studies." Someday in the near future you may be able to get your genome sequenced and find useful information that could guide your diet and improve your health. But that day has not yet arrived, and when someone says it has, make sure you ask to see the scientific studies on which the claim is based.

Preprescription Gene Tests

Should your doctor order genetic testing prior to prescribing a drug or suggesting a diet for you? Not yet. Unless there is a good reason to believe you might have a genetic profile that makes a specific drug dangerous for you to use, trial and error will continue to be used by most physicians when prescribing for most patients. And this is, as we have emphasized, unlikely to change until you and patients like you have your entire genome sequenced and saved in or linked to your electronic health record.

Pharmacogenomic biomarkers have been touted as providing "fantastic opportunities for personalized medicine," and we agree. If this potential were realized, it could have profound beneficial effects on our health care by improving drug efficacy, reducing costs, and most importantly, reducing adverse drug reactions. Adverse drug reactions, for example, account for about 7 percent of hospitalizations, 20 percent of readmissions to the hospital, and 100,000 deaths annually in the United States. On the other hand, most adverse reactions are the result of the wrong drug or the wrong dose, or drug interactions, not genetic variation. Matching drugs to genomes is not a magic solution to drug errors. Pharmaceutical and biotechnology companies have heavily invested in genomic-based strategies for developing new drugs, although their financial incentives may be different. Biotechnology companies primarily have sought ways to use genetic manipulation, including recombinant DNA methods, to make new drugs. This has, for example, been the major strategy of companies like Genentech, which produced recombinant human insulin, the first biotech product on the market. Genomic information is also expected to provide insights into the underlying biological mechanisms of disease and reveal biological targets and pathways that could lead to new drug discovery

It used to be argued that pharmaceutical companies are less interested in drugs designed for people with specific genetic sequences because this would fracture the market for their drugs, which are now sold for the entire population. Currently the opposite strategy is under way, as most major pharmaceutical companies see considerable profit in making expensive drugs for diseases that affect a small number of people. One example of this is Sanofi SA, the giant French pharmaceutical company, which recently purchased Genzyme, a U.S. company. Genzyme makes, among other products, a drug for people with rare enzyme deficiencies, such as Gaucher and Fabry diseases, which it sells for a very high price, making it unaffordable for people without health insurance.

Another example is Vertex, who with the financial support of the Cystic Fibrosis Foundation, produced the cystic fibrosis (CF) drug Kalydeco (ivacaftor). The drug, which the FDA commissioner touted as a sterling example of personalized medicine when it was approved in 2012, can only treat a very small percentage of CF patients (those with a specific genetic mutation)—about 1,200 in the United States. It is priced at more than $300,000 per year and needs to be taken for life. That means major profits for the drug company and healthy living for those lucky enough to get the drug. But as Barry Werth, the author of *The Billion Dollar Molecule*, asked in 2014, "How can we justify an eye-wateringly high-priced specialty drug for a few hundred patients that will cost as much as all other medications for everyone else with the disease combined?" He rightly worried not only about the cost to society of these specialized drugs, but also that working on rare diseases to develop similar specialized drugs means "less talent and money will be devoted to widespread scourges like diabetes" and tuberculosis. We applaud the development of a specialized drug market on behalf of the patients who will be helped. But pricing is a societal problem, and we must recognize that as long as specialized drugs are priced at exorbitant levels that only the rich and governments can afford, no matter how good and how genome-specific they are, they will serve primarily to make medicine even less affordable for society than it is today.

In addition to specialized drugs, other advances have occurred that translate pharmacogenomics into clinical practice. Today 10 percent of drug labels for FDA-approved drugs contain pharmacogenomics information. This information may describe variability in how individuals may respond to drugs, risks for adverse events, genotype-specific dosing, or information on how a particular drug works. Even among this small number of drugs where pharmacogenomic information has been approved for clinical practice guidance, controversies remain. A good example is clopidogrel (Plavix®), one of the most commonly prescribed drugs in the United States for prevention of heart attack and stroke in patients at increased risk for these problems. Clopidogrel works by preventing platelets from clumping together and forming blood clots. Uses of clopidogrel, particularly when combined with aspirin, include prevention of clot formation following placement of coronary artery stents and in a condition called acute coronary syndrome, a life-threatening form of coronary heart disease in which the heart muscle does not receive enough oxygen. This can occur during a heart attack or unstable angina when a person has severe chest pain, even at rest. Some individuals have specific variants in the *CYP2C19* gene that result in poor metabolism of clopidogrel that decreases its effectiveness in clot prevention.

The FDA approved a new label for clopidogrel in 2010 with a "boxed warning" about the diminished effectiveness of the standard drug dosing in individuals with these gene variants—that is, "poor metabolizers." As with warfarin, it was left to the clinician to decide whether genetic testing should be performed to guide therapeutic use of the drug. It also left vague how combinations of different variants of the gene should be taken into account ("intermediate metabolizers"). To make matters more confusing, two randomized trials suggest that there is no difference in outcomes of patients who have "poor metabolizer" variants and those who don't.

If your doctor were to start you on clopidogrel, would genetic testing for *CYP2C19* gene variants be useful in deciding the correct dosage? Unfortunately, "not yet" is still the right answer. This is because genetic information would likely be of only limited value. Why? First, in addition to at least twenty-five variants of the *CYP2C19* gene, variants in other genes, such as *ABCB1* and *CYP3A4*, have also been shown to affect the drug response to clopidogrel. Second, multiple other factors influence the effects of clopidogrel, such as drug-drug interactions (for instance, aspirin and proton pump inhibitors like omprazole, or Prilosec), smoking, diet (for example, caffeine intake), differences in platelet function tests, and diabetes. Third, for use in the emergency department, for instance when a person has an acute myocardial infarction (heart attack), either an existing genome in the medical record would have to be interrogated or a rapid and accurate genetic analysis would have to be performed. Neither is available today. Finally, further studies are needed to establish whether and which specific genetic variants affect clopidogrel response. On the other hand, in some cases gene testing prior to initiating drug treatment can be life saving. For example, *HLA-B*1502* gene variant testing is now required by the FDA prior to starting Asian patients on the antiepileptic drug carbamazepine, as described in Olivia's case in chapter 2.

Pharmacogenomics of the Future

Pharmacogenomics is only one of a series of new "-omics" technologies (e.g., proteomics, toxicogenomics, antibodyomics, infectomics) that will add to the ability of physicians to diagnose and characterize diseases, predict and improve drug efficacy, and reduce adverse drug reactions. These technologies have great potential if judiciously used in combination in the clinic. A letter to the editor in *Science* provides a reasonable caution: "The idealistic goal of personalized medicine and individualized drug therapy, which needs a holistic understanding of each individual patient's unique '-omics read-out,' is most likely unattainable—for the vast majority of complex traits—by advances in technology alone."

Much more work needs to be done before you can expect to see pharmacogenomics routinely incorporated into your health care or diet. This was well stated by Steven Nissen from the Cleveland Clinic Foundation (although we think he was wrong about BiDil, he remains a credible source on evidence-based medicine): "Unfortunately, in the popular press, the concept of personalized medicine has taken on a nearly cult like following with pronouncements describing how future physicians will use therapies that reflect the specific genetic makeup of individual patients. No matter how promising, pharmacogenomic approaches to treatment must withstand the same scrutiny required of all therapeutic advances—careful evaluation through well-designed randomized clinical trials."

This will require large numbers of individual research subjects willing to share their DNA sequence and medical records with researchers (on the scale of millions) in national and international collaborative efforts. How to decide whether to take part in such projects is a subject we explore in chapter 9. Much has been and is being learned, but we are still a long way from the day when we can use genomics to identify "the right drug in the right dose for the right patient."

Notwithstanding the current limitations and challenges of using pharmacogenomics in clinical practice, genomics has been revolutionary in cancer research. For generations the mainstay of cancer diagnosis has been based on microscopic appearance of tumors followed by hit-or-miss, one-size-fits-all approaches to treatment. However, as we discuss in some detail in chapter 8, increasingly we are seeing such traditional approaches being replaced by integration of genomics for diagnosis, prognosis, and targeted therapeutic options. It seems fair to conclude that for the immediate future most of the promise of pharmacogenomics will be for patients with serious, even potentially fatal, cancers.

To avoid the temptation of prematurely concluding that genomics really is much more about promises and predictions of success rather than actual accomplishments, we delay our exploration of cancer genomics to look at the beginning of life, where genomics has already made major changes that directly affect, among others, all pregnant women and their fetuses and children: reproductive medicine, including genetic screening of potential parents, embryos, fetuses, and newborns. Sometimes termed the reprogenomics revolution, reproductive genomics is the subject of the next three chapters.

WHEN THINKING ABOUT PHARMACOGENOMICS, CONSIDER THESE THOUGHTS

Racism and racial stereotyping have plagued genomics since its beginnings.

Do what you can to prevent genism.

Pharmacogenomics is in its infancy, and it is likely to grow only after your genome is digitalized and linked to your electronic health record.

It is unlikely that you or your physician will use genomics to determine your drug or dose in the near future.

Foods are analogous to drugs, and we are still learning about how our genes can determine how we react to foods and drugs.

CHAPTER 5

Reprogenomics

In the baby trade, demand is too high and desire too deep to be stopped. If parents want children and nature doesn't comply, then they are likely to pursue these children through whatever means possible. They will cross international boundaries and undergo untested, unregulated treatments.

—Debora L. Spar, *The Baby Business* (2006)

No other area of genomics has fostered more debate and more attempts at regulation than the new reproductive technologies, also known as reprogenomics. This is not only because reprogenomics centers on babies and pregnancy, but also because it is fostered by the creation of extracorporeal embryos and the private recruitment of "surrogate mothers" to produce children genetically related to at least one member of the contracting couple. In the United States, the reproductive medicine industry has proven impossible to regulate at the federal level and formidable to regulate at the state level. Social media, including Facebook and Twitter, have broken national boundaries and enabled a thriving international trade that the *Wall Street Journal* has described as providing customers with a way to assemble a "global baby." In their words, "With an international network of surrogate mothers and egg and sperm donors, a new industry is emerging to produce children on the cheap and outside the reach of restrictive laws."

A front page article in the Styles section of the *New York Times* in early 2014 featured a half-page photo of a gay couple (David Sigal and Brad Hoylman) with their three-year-old daughter Silvia. The child had been gestated by a woman who used a donor egg and sperm mixed from both men. It is not unusual for a gay couple to contract with a woman to bear them a child. The story was about New York State reconsidering its law against paying women to bear children—"commercial surrogacy"—and perhaps adopting a law more similar to California, which permits payment. Hoylman, a new member of the New York Senate, has put this issue on its legislative agenda.

The New York law banning commercial surrogacy was passed in response to the memorable New Jersey *Baby M* case. Mary Beth Whitehead had agreed to have a baby for a childless couple but changed her mind after the child was born, deciding she wanted to keep the newborn. The New Jersey Supreme Court ultimately ruled that the contract the couple had signed with Mary Beth, in which she promised them her child, violated the state's law against baby selling and was void. This transformed the dispute into a traditional custody case between parents. Custody was awarded to the sperm donor father because the judge thought it would be in the child's best interest to be raised by his father instead of his mother, who was given visitation rights. Mary Beth Whitehead was what is now referred to as a "traditional surrogate": she was artificially inseminated with the father's sperm, and her own egg provided the other half of the child's DNA. She was not only Baby M's gestational (birth) mother, but she was also her genetic mother. Surrogate motherhood today almost always utilizes a donor egg so that the surrogate mother is not genetically related to the child. It was this change that convinced the California Supreme Court that the "real" mother of the child in such an arrangement was the "contracting mother," the woman who hired the surrogate mother to gestate a child for her. In 2012 New Jersey Governor Chris Christie vetoed a bill that would have permitted some forms of payment to women who gestate children for others. Christie said he wanted more public discussion of "such a profound change in the traditional beginnings of a family."

In this chapter we return to the same issues that were so prominent in state legislatures a quarter century ago. What should you consider when deciding whether to support additional regulation of the new reproductive technologies, and what should you consider when comparing adoption to assisted reproduction methods that permit you to have a child with a genetic relationship to you or your partner? The nature of the fertility industry also matters. It is, as Debora Spar chronicles in *The Baby Business,* a highly commercialized sphere of medicine. It is also unique in that its product is a genetically related child, and the price the would-be parents are willing to pay "is literally inestimable." This is because if you seek a genetic tie to your child, "the possibility of substitution disappears."

Relationships and Reprogenomics

Genomics and reproduction are inextricably connected, and the transmission of messages from our genome is the basis for creating most family relationships. It is a biological imperative, sometimes referred to as a natural instinct, for all species (including humans) to survive, and the means of species survival is reproduction. This is a fundamental feature of life. Moreover, because children possess a genome made up half from the mother and half from the father, children are genetically distinct from their parents. This difference matters and is the basis of evolution, a change in the gene pool of a population over time. All evolution is based on genetic change. The environment "selects" those organisms best adapted to reproduce themselves. If environmental conditions change, organisms that possess the most adaptive characteristics for those new conditions will come to dominate because of their increased reproductive success. It is the mutations in the genomic endowment of individuals that enable such adaptive changes. Another way of saying this on the species level is "mutate or die."

Until relatively recently, human reproduction was almost exclusively based on sexual intercourse (the exception was the use of artificial insemination). With the advent, however, of what have come to be known as assisted reproductive technologies, or ARTs, it is now possible for humans to transmit their genes without sexual intercourse. ARTs are as remarkable as the sexual revolution brought about by effective contraception, which made sex without reproduction predictable. It should surprise no one that the social and ethical consequences of ARTs are every bit as revolutionary and remain the subject of intense public debate and speculation.

The birth of contemporary ARTs can be dated to the birth of Louise Brown in 1978. She is often inaccurately referred to as the world's "first test-tube baby" (figure 5.1). This description is inaccurate not because she wasn't the first (she was) but because there was no test tube involved. Her conception took place in a petri dish, and the embryo created (by combining her mother's egg and father's sperm) was transferred to her mother for gestation. The proper medical term for this procedure is *in vitro fertilization*, or IVF. Louise Brown's healthy

5.1 Robert Edwards with Louise Brown and her mother and son. Chris Radburn, "Professor Robert Edwards wins Nobel Prize," *Associated Press*, October 4, 2010.

birth came after more than a decade-long collaboration between an embryologist, Robert Edwards, and an obstetrician, Patrick Steptoe. The pair encountered intense opposition from religious, social, and even medical groups, both because of the implications of creating a human embryo in the laboratory and because of the world's overpopulation. As quoted in a *Science* article, people said, "Why would we help people having children when our main problem is how to prevent people from having children?" Others worried that the IVF procedure would create fetuses and infants with severe disabilities, and that there was no real science undergirding their "trial and error" methods. Criticism was silenced, however, with the birth of Louise Brown, who became an overnight international media sensation. Steptoe died in 1988. In 2010 Edwards was awarded the Nobel Prize in Medicine for his pioneering work, which opened the door to IVF and also to human embryo research and the derivation of embryonic stem cells. Had he survived, Steptoe would have shared the prize. By the time the prize was awarded, Louise Brown had given birth to her own child.

Nearly ten years after the birth of Louise Brown, in 1986, we published an article entitled "Social Policy Considerations in Noncoital Reproduction" in *JAMA*, the journal of the American Medical Association. We were so confident that it would quickly be recognized as the most scholarly and authoritative article on the subject that we purchased 1,000 reprints, anticipating requests for copies from around the globe (reprints are mostly a thing of the past, with PDF copies sent electronically today). A year later we still had about 950 reprints in our offices. "How could this be?" we asked ourselves. Maybe it was the timing, or maybe it was the title. The assisted reproductive technologies were still very novel and were known by a variety of terms, including "artificial reproduction" and "unnatural reproduction," but no one favored adopting our overly clinical term, "noncoital reproduction."

We believed it was prudent to reflect on the societal issues raised by these new techniques and arrangements, including IVF, use of embryos created in the laboratory, and surrogate motherhood. *People* magazine had just named Louise Brown one of the ten most prominent people of the decade, one who dominated it "by simply being." In the context of clinical medical practice, we predicted that it would "often be critical to make distinctions, previously irrelevant, between the genetic, gestational, and rearing parents when sorting out individual rights and responsibilities." A quarter of a century later, we can see that our predictions of confusion and conflict were, if anything, understated. A real-life experience from Sherman's hospital illustrates some still unresolved disputes.

Joan (not her real name) was transferred from a community hospital to Northwestern while experiencing vaginal bleeding and carrying twins at twenty-one weeks of pregnancy. Joan was a "surrogate mother" and was accompanied by the gamete donor couple (who had provided the egg and sperm, hiring her to gestate the twin pregnancy that she was carrying following IVF). The twins were both alive, but she had begun to dilate, and the membranes were protruding ("hourglassing") through the cervix into the vagina. A decision had to be made as to whether to try an emergency surgical procedure, called a circlage, in which the membranes would be pushed back into the uterus and a purse-string suture placed around the cervix to cinch it closed. It was explained to Joan and, at her request, to the gamete donor couple, that a circlage procedure carried risks to the fetuses, as well as to her. The likelihood of the procedure being successful in prolonging the pregnancy until the fetuses became viable was estimated to be low, in the range of 10–20 percent, but experience with cases like hers was very limited. Even if the pregnancy could be prolonged, there was a risk of delivering extremely premature infants who would require prolonged hospitalization in the newborn intensive care unit. The possibility of one or both infants suffering severe brain injury was very high. The risk to Joan was also high, particularly the risk of infection, which could lead to infertility or even a hysterectomy. The alternative approach was to induce labor with drugs and terminate the pregnancy.

Who should make the decision? Joan? The gamete donors (also the intended parents)? Joan's physician? A judge? The hospital's ethics committee? Basic principles of medical ethics and law give the authority to Joan because it is her body and she has the most at stake in the decision. When a medical intervention affects only the patient, the patient is afforded the legal right to make the decision. There is little room for dispute unless the patient is unable to understand the information necessary to make an informed choice. But what about this case, which involves not only the patient, Joan, but also two fetuses (to which she is genetically unrelated) and the genetic parents of the fetuses? It could be argued that since the entire purpose of the pregnancy was to provide the two gamete donors with genetic children, the architects of the "deal," sometimes referred to as a "family project," should be the primary decision makers.

It has been suggested that the gestational mother should have less to say about fetuses that she is carrying if they are not genetically related to her. The argument is that she should be considered not a pregnant woman at all but rather a type of container or vehicle, albeit a "viable vehicle," to deliver the fetuses to their genetic parents. This, however, seems wrong, because Joan is pregnant, is very much biologically (if not genetically) connected to the fetuses she is carrying, and is taking all the physical risk, and much of the psychological risk, of the pregnancy. Whatever the relationship between the fetuses and the gamete donors, the relationship between the fetuses and Joan is much closer and more intimate. We also know that Joan's body is the site of all decisions, and Joan is the only one who is guaranteed to be present at the time of birth to make decisions about the newborns' care if necessary.

All these reasons, we think, suggest that Joan should make the treatment decision. Joan's physician came to the same conclusion. The gamete (egg and sperm) donor couple fully supported Joan's decision to terminate the pregnancy, although they were involved in the decision making only because Joan permitted them to be. Had they disagreed with Joan and demanded that the pregnancy be continued, their demand should have been ignored. Within a few hours of inducing labor, the twins were born. Because they were previable, no resuscitation efforts were made, and they died. Following the extremely premature delivery, Joan experienced excessive uterine bleeding, and an emergency removal of the placenta was performed under general anesthesia. The procedure was complicated because the placentas did not separate easily from the wall of the uterus. Joan was given twelve units of blood and required admission to the intensive care unit for thirty-six hours. She was discharged three days later.

Few surrogate mother arrangements go so horribly wrong medically. But all such arrangements require us to identify both the mother of the fetus and resulting child, as well as the person who has the right to make treatment decisions regarding the fetus during the pregnancy and the child after birth. Before IVF, we never questioned the identity of the child's mother, even if the identity of the father was in doubt. Now both parental identities can be suspect, and it seems "natural" in our new genomic age to turn to DNA to define the mother-child relationship. This is often done, even though it seems simply wrong in a case like Joan's, where the biological contribution, as well as risk of life, is overwhelmingly on the side of the gestational mother and her relationship to the child. To date, uniformity in laws has proven impossible, and people commonly pick the state in which to engage a surrogate mother based on its particular laws (for instance, California has been particularly inviting, since it recognizes the contracting couple or "intended parents" as the child's legal parents). In the words of law professor Patricia Williams, "The [state] laws regarding surrogacy are a jumble of inconsistent public policies, free-market contract, civil interventions, and criminal sanctions."

Whatever state you are in, we don't think either genetics or contracts alone should determine parenthood and parental relationships. Rather we continue to believe, as we did at the dawn of the new reproductive technologies age in the 1980s, that the gestational mother (historically and accurately described simply as the "birth mother") should continue to be considered the child's legal mother for all purposes, with the rights and responsibilities to protect the child at birth. She can, should she choose so after birth, terminate her parental rights to the child. But termination of parental rights should always be her after-birth decision, not one imposed on her by contract or judicial decree prior to birth. Some commentators have suggested that the insistence that the birth mother continue to be legally considered the child's mother, even though she has no genetic relationship to the child, represents a refusal to acknowledge the new genetics and the "new families" genetic technology can help create. We disagree, because, we think, it is a violation of the gestational mother's human dignity to be treated as nothing more than an incubator. Moreover, we have noted that even as a matter of simple pragmatism in ART, the use of donor eggs (where the egg of a young woman is used to create an embryo for gestation by another woman—usually an older woman—who gestates and intends to raise the child) is much more common than the use of full surrogacy (where an egg from a donor—who could be the contracting woman—is used to create an embryo for gestation by a woman who intends to give up the child upon birth).

Applying the genetic model to donor eggs would mean that the woman who uses a donor egg to create an embryo that she gestates and gives birth to with the intention of rearing the child is not to be considered the child's mother (the egg donor is). This, we think, is socially and ethically unacceptable, as it puts all births in question: just delivering a child would no longer be sufficient to be presumed to be the child's mother. Instead, DNA samples would have to be taken and contracts examined to determine whose DNA was used, and therefore the child's genetic (real?) mother. Only a social policy that was indifferent to the rights and welfare of both pregnant women and children could, we think, support such a counterintuitive and genomic-controlling result. The "California solution" of using contracts to designate the "intended parents" as the real parents is equally unsatisfactory, because while it does identify the parents, it does so at the expense of making the child a market commodity whose family relationships are determined entirely by a commercial transaction memorialized in a contract.

Social Policy Issues and Reprogenomics

In our *JAMA* article, we discussed various assisted reproductive technologies, not just the use of a gestational or "surrogate" mother but also artificial insemination by donor (AID), IVF, and embryo freezing. To help prioritize the need to develop policies for these methods, we constructed a table in which we assigned a value for an "Index of Relative Importance of Societal Issues." The major issues considered were almost identical to the issues we continue to debate today: the potential of new methods to be put to uses other than infertility, ways to protect the human embryo, identification of the resulting child's mother and father, screening of gamete donors, donor anonymity, and opportunities for commercialization.

Many of these issues are still being actively debated, and there have also been some notable surprises. The primary surprise to us is how central commercialization and the market have become to ARTs, and how globalized the fertility industry generally has become. We gave no thought to the rise of the Internet and the ability of couples, gamete sellers, and potential surrogate mothers to use the Internet to find each other and make deals that did not directly involve physicians (or lawyers) at all. Commercialization has taken many procedures out of the hands of physicians altogether and placed them in settings devoid of any meaningful regulation or oversight—most notably the Internet, which has facilitated the globalization of ARTs. Couples now have the entire world to choose from for reproductive services. India, for example, has become a "go-to country" for hiring a surrogate mother, and an entire new industry has grown up around this practice there. Today no meaningful discussion of reprogenomics can ignore either globalization or the use of the Internet. Canada prohibits the purchase and sale of gametes and payment of surrogate mothers. This law has been upheld by their Supreme Court. Nonetheless, citizens of Canada can, and do, make commercial deals with women from the United States and India on a regular basis. The ARTs simply cannot be effectively regulated globally on a country-by-country basis nor in the United States on a state-by-state basis.

Not only have opportunities for commercialization increased, but so have the number of ART methods and societal arrangements, including (but by no means limited to) egg freezing, "oncofertility" (preserving fertility of children and adults undergoing treatment for cancer and other life-threatening diseases), preimplantation genetic diagnosis, embryonic stem cell research, use of "savior siblings" (having a child to harvest umbilical cord blood stem cells in order to save the life of another child with a life-threatening disease), and the most socially controversial practice: asexual reproductive cloning. Because this remains impossible, we will hold off discussion of cloning until the final chapter of the book. For defining relationships based on genetics, however, it is worth noting that asexual genetic replication would destabilize family relationships. A woman who used her own egg and somatic (body) cell to create an embryo, then gestated and gave birth to the child, would be simultaneously the child's mother and the child's sister. Another reason for postponing the cloning discussion is that cloning has little, if anything, to do with assisted reproduction—since whatever else one can say about cloning, there is no "natural instinct" or urge for humans to reproduce asexually.

The rights and welfare of resulting children have become more important in the ART realm, and it is worth noting the paradox that the rights and welfare of children have historically been a secondary consideration in this field. When we first wrote on this subject twenty-five years ago, we had a great deal to say about protecting gamete donors, explicating parental rights and duties, defining the status of embryos, and even dealing with commercialization. But we devoted only one sentence to what we argued should be the undisputable "foundation" upon which ART guidelines should be built: "Primary consideration should always be given to the 'best interests' of the potential child, rather than to the donors, the infertile couple, or the physician or clinic."

We continue to believe this, while acknowledging that this conclusion remains in some dispute and is not always easy to apply. It does answer the question of whether permanent records should be kept regarding the genetic parents (yes, because the child may want to know in the future) but does not so easily answer questions about whether the risks of specific procedures, like ICSI (intra-cytoplasmic sperm injection, injecting a single sperm into an egg, which could result in a child with the same genetic infertility problem as his father), are justifiable from the perspective of a child that would not have been born at all had ICSI (or any other ART with risks to the child) not been used. As Carl Djerassi has his character Frankenthaler note in his play about ICSI, *An Immaculate Misconception* (2000), "Before ICSI, men couldn't inherit infertility . . . it was uninheritable! But now?" It is at least a paradox that the infertility industry, dedicated to treating infertility, also creates infertile males—for whom it can supply additional treatment. Of course, simply ignoring risks to the child because it would not exist without running those risks proves too much: it would justify complete indifference to the welfare of future children. But requiring would-be parents to avoid any and all risks to future children is also too exacting a standard. A parental obligation to avoid unreasonable and predictable risks may be as good as we can do for now, while studies of the children of ART continue.

Finally, and perhaps most importantly, we noted that these new technologies "not only change what we can do with regard to human reproduction, but they also threaten to change how we think about human reproduction, and perhaps how we think about humanness itself." In the early days, for example, it was thought that the children of IVF, "test-tube babies," might be looked at (and look at themselves) as freaks. This has obviously not turned out to be the case. Instead, more and more women are postponing attempts to have a child until late in their thirties, using the existence of ARTs as a rationale. Major corporations, led by Apple and Facebook, now offer their female employees the "benefit" of egg freezing—an offer that is likely to further spur the nascent egg-freezing industry but may ultimately do professional women more harm than good by encouraging them to postpone childbirth until they have fulfilled their "obligations" to their employers. This is because postponing pregnancy can mean having no child at all. Egg freezing is not the biological "insurance " it is often marketed as—it is more in the nature of a high-stakes gamble that your eggs can be successfully used to create embryos that you will be able to carry to term. It is too early for meaningful data on the usefulness and effectiveness of egg freezing to postpone childbearing.

Simply because IVF and childbearing later in life have been mainstreamed does not mean that other methods of ART, including use of frozen eggs, will be equally safe, effective, or accepted. In fact, as we observed in our *JAMA* article, IVF using the gametes of the married couple actually presents almost no new issues other than the creation of an extracorporeal embryo—that is, an embryo created outside the human body. This issue, however, is not a relationship one but rather a research one: the very existence of extracorporeal embryos has enticed scientists to manipulate and utilize the embryo for nonreproductive purposes, including the creation of embryonic stem cells.

Additional screening of extracorporeal embryos in IVF can also be seen as an integral part of IVF itself. Termed preimplantation genetic diagnosis, or PGD, the procedure combines IVF with manipulation of the embryo to remove a cell from it five to six days after fertilization. The cell's genome is then examined to determine the presence or absence of specific genetic traits in the embryo. If the embryo is found genetically acceptable, it is used to attempt a pregnancy. PGD has been used primarily to select embryos that do not have conditions such as cystic fibrosis or myotonic dystrophy. By identifying embryos with aneuploidy and not using them to attempt a pregnancy, the success rate of IVF has been dramatically improved. The most controversial use of PGD is to use the technique to either screen for late-onset disease (diseases that won't manifest until adulthood, if at all) or for specific traits, such as gender, or even hair or eye color. Looking for such characteristics has been labeled "designer babies," and this seems a fair description of the goal. Screening for serious genetic diseases is one thing; screening for social reasons is another, and it should be discouraged if not outright condemned, because it directly commodifies children by valuing them not for themselves but for specific "desirable" characteristics.

PGD can also be used to prevent the use of an embryo with a serious genetic condition, and even keep the knowledge of the detrimental condition from the woman. Barbara Bailey (not her real name), for example, was worried that she might pass a gene (*PSEN1*) that caused early Alzheimer disease in her mother to her child but didn't want to know whether she herself had the gene. Genomic practitioners were able to help her. Using in vitro fertilization (IVF), Barbara and her husband created embryos in a petri dish. In similar fashion to the technique used in the movie *Gattaca*, preimplantation genetic diagnosis was used to screen each embryo to see if it carried the *PSEN1* gene. Two embryos that did not have the gene were then implanted into Barbara Bailey. Barbara herself was able to be certain her children would not have the *PSEN1* gene without simultaneously learning if she had it herself. There are some genomic messages we want to hear, and some genomic messages we may not want to know about.

Related to PGD is a technique that conjures up even more anxiety about creating designer babies: so-called three-parent IVF, or mitochondrial manipulation techniques. Mitochondria are small specialized structures found in the cytoplasm that act as tiny power plants to provide energy essential for cellular function. Power-hungry cells in such tissues as brain, heart, muscle, liver, and kidneys contain the most mitochondria—up to several thousand per cell. Mitochondria have their own genomes separate from the genome present in the nucleus of the cell; mitochondrial DNA (mtDNA) are circular and contain only thirty-seven genes. Mitochondria are almost exclusively inherited from the mother; the egg usually destroys mitochondria in sperm during fertilization. About 4,000 children are born in the United States annually with a mitochondrial disease.

A technique has been developed that could allow women carrying mtDNA mutations to circumvent passing on a mitochondrial disease to their children by replacing mutant mtDNA with normal donor mitochondria in unfertilized eggs or at the single-cell-embryo stage. Another rationale for using the technique is to increase the likelihood that the egg from an older woman could be successfully used in IVF (by transferring the cytoplasm from the egg of a younger woman to the older woman's eggs). The feasibility of treatment of mtDNA diseases has been demonstrated in animal models and in human eggs, and the FDA is currently deciding whether it is safe to permit research on humans with this technique (the FDA is not considering the ethical issues). The primary ethical arguments against using the technique are that it crosses the "bright line" of germ line genetic modifications—genetic modifications producing "designer babies" whose design genes will be passed down to future generations and could harm them—and that given the unknown risks of the procedure, alternatives (including use of a donor egg or PGD) should be pursued instead.

The designer baby argument is formidable, and we will return to it in chapter 10. A child could be considered genetically designed by modifying only one gene. The "three-parent baby" argument, on the other hand, is much weaker because supplying 37 genes out of 20,000 is hardly sufficient to be named a third parent, any more than supplying a single replacement gene—were this possible—makes the source of the replacement gene a parent. The central question then remains: do we have a compelling reason to cross the line into genetic engineering? The argument is a frail one, and there seems no reason at all to cross the line for anyone who supports the use of IVF with PGD or donor eggs. In early 2015 the House of Commons, by a vote of 382 to 128, approved the licensing of facilities to begin to use this procedure in England. Approval by a publicly-elected body, with oversight by a public agency, seems just right to us as a way to introduce extremely novel procedures into clinical trials. On the other hand, without any public debate, there are scientists who argue that this decision gives the green light to tinker "with the rest of our genome." This is a serious misreading of the vote.

Extracorporeal Embryos and Stem Cell Research

Manipulating the human embryo for reproduction purposes is controversial because of its effect on both the resulting child and society's views on parenthood. But any manipulation of human embryos remains highly controversial, even if the embryos are not destined to be used for reproduction. The use of human embryos for medical research, for example, has remained the most politically contentious medical research area since IVF was introduced. Debate over federal funding of such research continues to this day, with the only hope of compromise centered on using so-called leftover or spare embryos created for use in IVF clinics. This is because such embryos were created for a "legitimate" reproductive purpose (not to be used instrumentally, but simply as means to a research end), and when the couple no longer wishes to or needs to use these (now frozen) embryos for reproduction, it seems at least as legitimate to donate them for research as it does to destroy them or leave them frozen forever. One of the people we acknowledged in our 1986 *JAMA* article was peer reviewer Leon Kass. He did an exceptionally careful review of our submission, and we were grateful. Kass later became, we believe, overly fixated on human embryos. He was one of two ethicists who helped persuade President

George Bush to limit federal funding to stem cells obtained from IVF embryos *before* his August 9, 2001, speech to the nation on this subject. In that speech, Kass was also named to head Bush's Presidential Bioethics Commission.

It is a very minor footnote that George was asked to testify before this commission on ARTs in 2003. The major points George made were that the industry could not regulate itself (we had both concluded this on the basis of the more than five years we spent together on the Ethics Committee of the American Society of Reproductive Medicine), and the old "best interest of the sperm donor" model needed to be replaced with a child-centered, best interest of the child model. George also suggested that the commission had to make a fundamental decision about either concentrating their work on embryo research (an area in which consensus continued to seem unlikely) or broadening their agenda to suggest ways to regulate the ART industry. Unfortunately, we think, the commission chose the former. We were less kind to the commission and its agenda, which seemed to us and many others to have become overtly political, in a commentary we wrote for *Nature* that was published during the 2004 Democratic convention. We asked, "Can bioethics in the US rise above politics?" We didn't think it likely, noting of the Bush/Kass bioethics commission: "Born with an embryo-centric, anti-abortion and anti-regulation political agenda, Bush's President's Council on Bioethics has repeatedly failed to transcend it." We continued, with specific reference to ARTs, "In its report, *Reproduction and Responsibility*, the council attempts to come to grips with the ethics of assisted-reproduction technology, but ultimately reverts to embryo protection. A thoughtful national report on public oversight for the assisted-reproduction industry is long overdue in the United States."

Federal funding for human embryonic stem cell research remained the exclusive ART-related issue debated at the federal level during the Bush administration, and this remains the exclusive issue in the Obama administration. Near the end of the Clinton administration, the National Institutes of Health (NIH) took the position that federal funding could be used to do research on stem cells derived from human embryos so long as no NIH funds were used to "destroy" the embryos, and the embryos themselves were not created for destruction but were "leftover" or surplus embryos created for reproductive purposes via IVF and were given for research purposes with the informed consent of the couple. Bush limited the application of this rule to stem cell lines created before his August 2001 speech on the subject. Obama reverted to the Clinton rules. His reversion was challenged in court, but the challenge was unimpressive and ultimately unsuccessful. Current NIH rules, which will remain in effect at least until Congress acts (an extremely unlikely prospect), are that federal funding can be used for research on human embryonic stem cell lines created by private funds. This makes no sense from an ethical point of view, but only from a pragmatic political perspective. The moral status of the human embryo no more depends on how it was created than the moral status of a child depends on how it was created, including whether it was the product of ART or even cloning.

Regulation of the Fertility Industry

The fertility industry is big business. There are more than 450 fertility clinics in the United States alone, and their ART procedures result in about 50,000 live birth deliveries and 62,000 infants a year. This is about 1.5 percent of all live births but 5.7 percent of all low-birth-weight babies. Of ART babies, 36 percent are born premature (compared to 12 percent non-ART babies). Most impressive (and a major downside of IVF) is the high percentage of multiple births, which can put both mother and babies at risk. Singletons account for about 55 percent of all ART infants, with 42 percent twins and 3 percent triplets or more. This compares to 97 percent of all non-ART births that are singletons. The incredible incidence of multiple births and premature infants in ART is a problem that has yet to be effectively addressed, although transferring single embryos seems likely to be the primary solution. A market approach would be to require ART clinics to pay for the neonatal ICU care of multiple births. This would discourage use of more than a single embryo per cycle.

The specialty of infertility medicine is based on, and often justifies itself by referring to, the deep desire of individuals to perpetuate themselves—and to most of us that means perpetuating our genes. It can be argued that genetic replication is the only way we (or at least our genes) can achieve immortality. This is also why fertility and genetics are inextricably linked, not only in a biological sense but also psychologically. It helps explain why many go to such great lengths in pursuing ART to achieve genetic parenting.

The intense desire for a genetically related child may also go a long way to explaining three of the most bizarre cases in the infertility industry: Cecil Jacobson, Ricardo Asch, and Michael Kamrava. We summarize these cases not because they are typical—thankfully they are not. They do, however, illustrate that in the absence of any effective regulation of reprogenetics, patients can be left at the mercy of the market and unscrupulous practitioners, and that the industry does not, and perhaps simply cannot, police itself. "Buyer beware" may be a good slogan at a department store, but it should have no place in the practice of medicine. The cases also suggest additional important issues in assisted reproduction, the kinds of regulation that might effectively protect infertile patients, and things you may want to consider in picking an infertility clinic. All three of these unusual cases are scandalous events in the infertility industry in general, and in the field of infertility medicine in particular. Patients were betrayed. Physicians went far beyond malpractice to engage in fraud. Their unethical actions affected not only the patients involved but also their genetic progeny.

We begin with Cecil Jacobson, a physician who misused the trust patients placed in him by artificially inseminating his infertile patients with his own sperm. In the 1960s, Cecil Jacobson received international notoriety as a pioneer in the development of genetic amniocentesis when he was on the faculty at George Washington University. He later established an infertility clinic in Fairfax, Virginia. In 1989, a former patient complained about his clinical practices. During an investigation, it was discovered that Jacobson had artificially inseminated dozens of patients with his own sperm, telling them that the sperm had come from anonymous donors. DNA tests linked Jacobson to at least fifteen children "fathered" by his sperm. He was sentenced to five years in prison for the federal crime of mail fraud, and his license to practice medicine was revoked (George was a consultant to the federal prosecutor). After he served his sentence, Jacobson moved to Provo, Utah, where he is said to be involved in agricultural research.

What is unusual about the Jacobson case, of course, is that it was the physician who was fixated on perpetuating his own genes, even at the expense of the rights and welfare of his patients. Taking unconscionable advantage of his patients and using his own sperm for artificial insemination (AI), he fathered perhaps one hundred children. We can ask what his social relationship is with the children (hopefully nonexistent), and perhaps more important, what their relationship should be to each other, since they are all half siblings. There was some concern that when the children grew up they might meet and marry each other, creating a higher risk of genetically compromised children. How should this be dealt with? And more generally, should there be a registry of all children conceived by donor sperm or donor egg so they can access their records when they turn eighteen to "discover" their genetic father or mother as well as any half-siblings? Developments in social networking and direct-to-consumer genetics companies have made it more feasible to identify genetic relatives. It is not possible to regulate contact via the Internet, and even if it was, who should set the rules for contact?

Sperm donors have been the model for policy making in ART (for example, "If men can sell their sperm, women should be able to sell their eggs," and "Genes should determine parenthood"), but should the male model apply to egg donors? This question can be considered in the context of another exceptional case, that of Ricardo Asch. A native of Argentina, Asch came to the United States, where he developed a new reproductive technique called "gamete intrafallopian transfer," or GIFT. GIFT is a variation on in vitro fertilization: eggs are removed from a woman's ovaries and placed in one of her fallopian tubes, along with the man's sperm, allowing fertilization to take place inside the woman's body. The Pope John Center, in consultation with the head of the Pope Paul II Institute for the Family in Rome, approved the GIFT procedure in 1985, as long as the sperm are collected during sexual intercourse. In 1984 Asch was recruited to the University of California at Irvine, and in 1990 he was named director of the Center for Reproductive Health. In 1992, staff members of Asch's fertility clinic began to suspect that embryos were being "misappropriated." Ultimately Asch and two of his colleagues were accused of implanting "extra" eggs that were fertilized in vitro into as many as forty unknowing women, resulting in the birth of at least eight children who were genetically unrelated to the birth mother. In 1996, one of his colleagues was found guilty of mail fraud involving illegal billing. He was sentenced to three years probation and fired from the university. Another colleague fled to Chile, where he now practices medicine. The University of California paid more than $24 million for more than 130 incidents in which eggs or embryos were either implanted into nongenetically-related women or could not be accounted for. Asch fled the country to practice medicine in Buenos Aires. It was reported in 2014 that he was arrested in Mexico and later released after the Mexican authorities decided not to extradite him to face criminal charges in California.

Asch later further outraged his victims by dismissing the importance of genetics to reproduction. In an interview with Diane Sawyer on ABC's *Prime Time Live,* Asch blamed Americans for being overly concerned about their genes: "I think it's [a] much more materialistic society than other societies, the craziness about genes. . . . I think it's entirely obsessed, this society, with genes." He said genes "are, at least in my opinion, not that important. . . . I know they're not important for IQ, for athletic ability." This is a remarkably self-serving (and hypocritical) statement from an infertility physician who makes almost all his money by marketing precisely the opposite view of genes.

The final case is generally referred to by the name of the patient rather than her fertility physician, Michael Kamrava. Nadya Suleman is known around the world as the "octomom," because she had eight babies in one IVF pregnancy (she reportedly insisted that her physician transfer all six embryos that had been created for IVF, two of which split). That really is all that has to be said about her case. Although some fertility experts, like Jamie Grifo, have argued that physicians should have nothing to say about how many babies women have, this position seems to us to be simply an abdication of responsibility. The fact that a patient can refuse any treatment does not mean that a patient, in the words a New Jersey court, can "demand mistreatment." Nadya Suleman's eight-baby IVF pregnancy was medical malpractice per se, an extraordinarily dangerous and unnecessary pregnancy not only for Ms. Suleman but also for her eight fetuses. It was appropriate that Kamrava lost his California medical license for this and other subpar care. It is also worth noting that although "infertile" by the industry's definition (unable to become pregnant after a year of sexual intercourse), she already had six IVF babies and now has fourteen children. To her credit, we think, she became part of a PETA advertising campaign to encourage pet owners to spay or neuter their pets (figure 5.2).

5.2 Nadya Suleman endorsing PETA's advertising campaign. Krista Kennell, "Octomom unveils PETA sign," *Associated Press*, May 19, 2010.

Infertility medicine can rightfully be viewed as a subset of genomic medicine—here, not using your genes to help design a treatment or prevention program, but using your genes (as packaged in eggs and sperm) in a way that permits you to have genetically related children, your "personal" (genomic) children. Infertility medicine permits the combination of an individual's genome with the genome of the other genetic parent. To enable one's genome to survive is the goal of reproduction, making infertility medicine a subset of genomic medicine. Of course, if genomic reproduction were not the objective of infertility treatment, adoption would be the surest, safest, most cost-effective, and most socially responsible alternative.

Great strides have been made in ART over the past three decades, but we still have a long way to go. Most important, we think, is the development of consistent best practice guidelines by the infertility industry, backed up by consistent state laws that are designed primarily to protect children. Ultimately we will need international treaties to protect all parties involved in ARTs when people from outside the couple are engaged to help them create a genetically related child that they can rear. This is consistent with treaties for the protection of adopted children. Until such laws and practices are in place, couples looking to hire others to help them have children must proceed with extreme caution under a rubric usually not used in medicine: buyer beware.

Combining the fields of ART and genetics first led to preimplantation genetic diagnosis, which for couples at high risk allowed selection of embryos free of a specific disorder, such as Tay-Sachs disease or cystic fibrosis. In an editorial in the journal *Fertility and Sterility*, Richard S. Scott and Nathan R. Treff describe how rigorous evaluation will be needed as new genomic technologies change how embryos are screened to optimize the success of achieving a pregnancy and delivering a single healthy baby. Put another way, only those embryos with the greatest potential based on genomic testing will be selected for transfer. Presumably, this will allow the ideal of a single IVF embryo to be transferred without reducing pregnancy rates. (Twins could still occur, however, because a single transferred embryo can split.) The process of determining which embryo is "the most fit" for transfer is scientifically complicated and will require scientific research, including randomized trials, to assess both the safety and efficacy of any test.

At the beginning of this chapter, we noted that most family relationships are defined by genomic messages. We end with the acknowledgment that ARTs have made genomics much less determinative in defining relationships (think, for example, about egg donors). Nonetheless, even if not always determinative, genomics does provide an important way to view this question. Of course, genomics is inherently personal in reproduction that involves the transmission of our genome to our children. But genomes can also be viewed as universal—as geneticist Aravinda Chakravarti has put it, "All living humans are mosaics with ancestry from the many parts of the globe through which our ancestors trekked. In other words, each of us has around 6.7 billion relatives." In this sense, the fertility industry's crossing of international boundaries could help usher in not only a global business model for ARTs but also a better global understanding of universal human rights and human dignity. In the next chapter, we continue our discussion of the role of genomics in reproduction by shifting our focus to the human fetus.

WHEN THINKING ABOUT REPROGENOMICS, CONSIDER THESE THOUGHTS

The interests of the one unrepresented party to ART, the child, should be of primary concern to all parties.

"Surrogate" mothers are real mothers and should be treated with dignity.

Sperm donor rules should not dictate egg donor rules, because the risks are not the same.

Multiple IVF births can only be prevented by limiting the number of embryos transferred.

Reprogenomics is a massive unregulated commercial industry whose clients would benefit from either enforceable best practice standards set by physicians or federal or global regulation.

Reprogenomics is a global unregulated business that makes it possible for you to bargain for a genetically related baby, but you need to follow basic "buyer beware" rules to protect yourself and your child from exploitation.

CHAPTER 6

Genomic Messages from Fetuses

If disclosure and consent are ever to have meaning, physicians must learn to manage uncertainty better.

—Jay Katz, *The Silent World of Doctor and Patient* (1984)

The increasing ability of science to prenatally diagnose genetic disorders and congenital anomalies in the fetus is the subject of this chapter. Progress has proceeded so fast in the past decade that there is little doubt the technologies we summarize will have evolved further by the time this book is published. Nonetheless, the social, legal, ethical, and public policy issues raised by increasing the number and types of genetically determined or genetically influenced traits that can be detected in the fetus will not change and are ultimately more important than the science in determining what you can and should do with this newly acquirable genomic information. The opening quotation from Jay Katz, the world's leading authority on informed consent, is meant to suggest that even with all our new technologies, uncertainty is still at the heart of decision making during pregnancy.

As you read this chapter, envision yourself as pregnant and ask yourself the following questions: What are the "boundaries," if any, that should restrict or prohibit prenatal diagnosis (for example, the severity of the condition for which testing is performed, stage of pregnancy, likely planned actions if an abnormality is detected, use to determine fetal sex or other nondisease characteristics, such as eye color, height, athletic ability, musical talent, or intelligence)? When you see your physician and he or she offers a prenatal genetic test, do you want to be told simply that your physician "recommends" the test, or would you rather have your physician explain the benefits and risks of the test in a "neutral" (unbiased) manner and leave the decision of whether to have the test strictly up to you?

Other questions you should be thinking about involve the level of detail. How much information do you need to make an "informed decision" about prenatal genetic diagnosis? Do you want to know about all the symptoms, treatments (if any), underlying causes, and prognosis of each disorder? Would you want to have this information before the test is performed, or would you rather wait until the test results are back, dealing with these issues only if they are relevant to you or your family? Do you want to know the likelihood that the test will show a problem in the fetus and the accuracy of the test? If you are told that the fetus has a serious genetic abnormality or anomaly, what would be the best course of action for you to take (that is, continuing or terminating the pregnancy), and who should make this decision? What additional information would you want from your physician? Would you go to the Internet before deciding? Do you think the more information you have the better? These are all difficult questions, and your answers will depend not only on medical factors but also on your personal beliefs and values.

In 2013 the American College of Medical Genetics and Genomics (ACMG) opposed legal restrictions on abortion following prenatal diagnosis. The college noted that the entire practice of medical genetics is providing patients with information to enable the pregnant couple "to choose a safe and personally acceptable management plan," and concluded that "termination of pregnancy for genetic disorders or congenital anomalies that may be diagnosed prenatally is a critically important option." This statement was prompted by the passage of a North Dakota law outlawing abortion for a "genetic abnormality," defined as "any defect, disease or disorder that is inherited genetically." The law also has its own nonexclusive list of prohibited indications for abortion: "Any physical disfigurement, scoliosis, dwarfism, Down syndrome, albinism, amelia, or any other type of physical or mental disability, abnormality or disease." This law and proposals like it are a direct response to the increasing number of genetic tests that can be performed on fetuses—and thus the increasing number of conditions that might lead pregnant women to terminate their pregnancies. North Dakota governor Jack Dalrymple said he signed this law to challenge the "boundaries of *Roe v. Wade*," the 1973 U.S. Supreme Court opinion on abortion.

Roe provides that a pregnant woman, in consultation with her physician, has a constitutional right to terminate her pregnancy prior to fetal viability, and that she retains that right even after fetal viability if her "life or health" is endangered by continuing the pregnancy. The state may impose some requirements on previability abortions, such as requiring informed consent and a twenty-four-hour waiting period, but only if these conditions do not "unduly burden" women by actually preventing them from obtaining abortions. The chances that even the current Supreme Court would find the North Dakota law constitutional approach zero because the law is so extreme. For example, it would prohibit abortions even in cases where the fetus is nonviable because of anencephaly (absence of a major part of the brain) and cases where the baby would die a slow, painful, and inevitable death, such as Tay-Sachs disease. Nonetheless, just as society is unsympathetic to abortion when the fetus is the "wrong" sex, we believe society is likely to continue to support "genetic abortion" only if it is limited to serious genetic conditions. Down syndrome has historically been considered the classic serious condition, and as a descriptive matter but not an ethical one, new conditions are often looked at as being more or less serious than Down syndrome. As for Down syndrome itself, 60–90 percent (the number varies by survey and depends on a number of factors) of women whose fetus is diagnosed with Down syndrome elect to terminate the pregnancy.

Society has not had to decide what genetic conditions justify abortion, but as we will discuss, the rapid expansion of genetic tests of the fetus makes this a compelling contemporary quandary. Can we simultaneously reject eugenics and discrimination, while providing pregnant women with ever more genomic information about their fetuses? Before we can begin to answer this question, it is critical to understand how we got to this point and what values underlie the practice of prenatal diagnosis.

Prenatal Diagnosis

Prenatal genetic diagnosis began in the mid-1960s with the development of amniocentesis, a procedure in which a small sample of amniotic fluid surrounding the fetus is withdrawn from the uterus via a needle inserted through the pregnant woman's abdomen. Either the fluid or the cells suspended in it are used to detect serious fetal genetic abnormalities. Among the first to perform amniocentesis in clinical practice was Albert Gerbie at Northwestern University, Sherman's early mentor and the obstetrician who delivered Sherman's two sons. In the beginning, amniocentesis was considered far too dangerous because it was thought to increase the risk of miscarriage and serious complications in the pregnant woman.

Gerbie and others who pioneered prenatal genetic diagnosis using amniocentesis did so at considerable peril to their own careers. Research on amniocentesis occurred before *Roe v. Wade,* at a time when in most states women had great difficulty in obtaining authorization for an abortion, even when a serious fetal abnormality was diagnosed. Twenty years after *Roe,* a committee of the Institute of Medicine (IOM) suggested clinical guidelines for prenatal diagnosis. The committee prophetically anticipated that many more genetic tests would be developed, and "eventually technologies will be available to simultaneously test for hundreds of different disease-causing mutations, either in the same or different genes." The committee's recommendations, which remain valid today, included these:

- Patients must be fully informed about the risks and benefits of testing procedures, their possible outcomes, and alternatives.

- Prenatal diagnosis should only be offered for the diagnosis of genetic disorders and birth defects, not for minor conditions or characteristics, or for fetal sex selection.
- Education before and after prenatal screening should be available to patients, and ongoing counseling should be available following pregnancy termination.
- Reproductive genetic services should not be used for the eugenic goal of "improving" the human species.

The recommendations became the foundation upon which professional organizations, most importantly the American College of Obstetricians and Gynecologists (ACOG) and the American College of Medical Genetics and Genomics (ACMG), built their own guidelines for prenatal diagnosis. The amount of genomic information that can be obtained from fetuses, as well as the variability and uncertainty of genetic findings, is so great that it is often difficult for anyone to interpret their meaning. Prenatal diagnosis can, however, sometimes identify a serious condition, and this information could lead a couple to terminate a pregnancy. The ability to determine the health of the fetus also permits couples that would not otherwise take a chance on having a baby because of the high risk of a serious genetic condition, such as Tay-Sachs disease, to have children without risking the birth of an affected child.

Genomic information about fetuses can also include information that is unclear and even contradictory. Such information can increase uncertainty and result in confusion, anxiety, and difficult decision making for everyone—physicians, patients, and families alike. In a debate on the future of the new genetics with Francis Collins at a national genetics meeting in New Orleans in 1993, George (who was asked to address the "dark side" of the Human Genome Project) suggested that the proliferation of genetic and genomic information about fetuses would, at least temporarily, lead to a marked increase in abortions. Geneticist Paul Billings had made this suggestion a year before, saying, "For now and for the foreseeable future, a major benefit derived from genetic information by families and individuals is the possibility to prevent the birth of other gene carriers by utilizing selection abortion." At the same congressional hearing in which Billings made this suggestion, the then head of the National Institutes of Health (NIH), Bernadine Healy, said she did not think the genome project was directed at abortion. Collins, who is currently the head of NIH, would undoubtedly take the same public position. Even in 1993, Collins was not happy to hear this suggestion, and both he and Healy were certainly correct in suggesting that increasing the number of abortions was not the goal of the Human Genome Project. Nonetheless, for the foreseeable future the new genetics will be used primarily to enable much more detailed prenatal screening and diagnosis, and will therefore increase the number of abortions. Preventing these abortions is the stated goal of laws like the one passed by the North Dakota legislature.

Most commentators and many physicians use the terms *screening* and *diagnostic testing* interchangeably. *Screening* usually refers to looking for problems or conditions in the general (healthy and asymptomatic) population, whereas a diagnostic test or testing is used to see if someone suspected of having a disease or condition actually has it. Examples of screening the fetus during pregnancy include a combination of maternal serum biochemical markers and ultrasound findings to determine if a fetus is at increased risk of having Down syndrome. A diagnostic test, on the other hand, identifies individuals or fetuses that actually have a specific disease or condition; it diagnoses a condition. There are approximately 4 million births annually in the United States. All these pregnancies are candidates for prenatal screening and diagnosis. Our best estimate is that about 350,000 pregnant women undergo prenatal diagnosis annually in the United States, and about 70 percent of all pregnant women undergo screening for fetal Down syndrome. It is worth stressing that all these tests, both screening and diagnostic, are voluntary and should never be performed without the informed consent of the pregnant woman.

Vanessa and Steve (not their real names) were treated by Sherman. We tell their story because it illustrates an inherent paradox in modern obstetrics: in trying to make pregnancy and childbirth safer for women and increase the probability of having a healthy baby, it is necessary to convey information (and perform tests) that will likely significantly increase the anxiety of prospective parents. Some physicians tend to go overboard and give an exhaustive list of all possible problems just in case lightning strikes; this is often referred to as "defensive medicine." The thinking is that it is better to "overinform" just in case something unanticipated goes wrong, so the physician cannot be blamed for failing to mention the possibility. Of course, simply increasing the amount of information conveyed risks confusing the patient, losing perspective of the most important and common issues that should be considered, and inflicting needless anxiety. How much information someone really needs to be an informed patient is not an easy question. Each patient really is different, and there is no "right" answer that applies to all patients. Informed consent has basic minimums but ultimately should also be personalized. The minimum is that all patients must be provided with information that might lead a "reasonable patient" to reject the screening, but information should be tailored to the individual's risks, including the health of prior children and the ethnicity of the couple.

As you read the story of Vanessa and Steve, ask yourself if you think Sherman struck the right balance between giving too much or too little information. Would you have wanted more or less information than Sherman provided? Would Sherman's use of more simplified language have been condescending, or was his expectation of the couple's medical knowledge too great?

Vanessa, a twenty-nine-year-old social worker, and her husband Steve, a thirty-two-year-old electrical engineer, were very excited; they had just learned that they were going to have a baby, their first. When Vanessa was about ten weeks pregnant, Sherman saw her and Steve at her first office visit. Sherman began with a general discussion of the dangers of cigarette smoking, alcohol and drug use, and excessive weight gain during pregnancy, as well as the importance of prenatal vitamins and proper diet, and he recommended an influenza vaccination. He told the couple that the laboratory tests he routinely obtains on all prenatal patients include a Pap smear, blood type, Rh factor and antibody screen, a one-hour glucose challenge test for diabetes, urine culture and protein assessment, and screening for infections, including syphilis, gonorrhea, chlamydia, hepatitis B, Group B strep, and HIV. He also told them that most women have several ultrasounds during pregnancy, and that this was useful in determining the gestational age, the number of fetuses, fetal viability, the location of the placenta, fetal growth disturbances, and amniotic fluid volume, as well as detecting major fetal anomalies.

Vanessa and Steve completed a detailed intake questionnaire that included a family history of genetic disorders and birth defects. Sherman explained that the American College of Obstetricians and Gynecologists recommends that invasive prenatal testing—first-trimester CVS or second-trimester amniocentesis—for chromosomal abnormalities be made available to all women, regardless of age (the older guideline was to test at age thirty-five). He explained that amniocentesis involves inserting a needle into the uterus and amnion (the sac surrounding the fetus) under ultrasound guidance and withdrawing a small amount of amniotic fluid. Chorionic villus sampling (CVS), which is also performed under direct ultrasound visualization, entails collecting a small sample of placental tissue either by inserting a thin flexible tube through the vagina and cervix into the uterus or by inserting a needle through the woman's abdomen. Both CVS and amniocentesis carry small risks of complications (less than 0.5 percent), most importantly miscarriage, infection, leakage of amniotic fluid, or vaginal bleeding after the procedure. Fetal injuries are very rare.

Sherman went on to say that various noninvasive first- and second-trimester fetal screening tests (not diagnostic) can be used rather than going directly to invasive prenatal testing. Specifically, to achieve a higher rate of Down syndrome detection (and a few other chromosome abnormalities) and reduce the need for invasive tests, an "integrated screening" approach can be used. This approach combines first-trimester ultrasound measurement of a fluid-filled space behind the fetal neck, called the nuchal translucency, with maternal blood biochemical markers. These first-trimester screening tests are then followed by additional maternal blood markers, collectively called a "Quad test," performed in the second trimester. These tests can detect about 95 percent of all fetuses with Down syndrome.

An alternative to integrated screening would be for Vanessa and Steve to choose between two "sequential screening" strategies. Sherman said that his own preference is the "stepwise sequential strategy." If the risk for Down syndrome was greater than 1 in 30, based on a first-trimester serum screening and nuchal translucency measurements, Vanessa would have a second-trimester Quad test, after which the results of both tests would be combined. She and Steve could then decide (based on the risk of Down syndrome) if they wanted an amniocentesis. The second approach is "contingent sequential screening." Vanessa would be assigned to one of three Down syndrome risk groups based on the first-trimester screen results. If the risk was greater than 1 in 30, CVS or amniocentesis would be performed; if the risk for Down syndrome was "intermediate" (between 1 in 30 and 1 in 1,500), second-trimester screening would be performed. If the risk was lower than 1 in 1,500, no further testing would be done. This approach lays out a plan at the outset and lets the couple avoid having to make a series of decisions as the pregnancy proceeds.

Sherman also said that during the second trimester a blood test, called maternal serum measurement alpha-fetoprotein, or MSAFP, is routinely offered as a screening test for certain neural tube defects. The most common neural tube defects are spina bifida (in which the spine is partially open, which can lead to nerve damage and varying degrees of disability) and anencephaly (in which a large part of the brain and skull fail to develop, and which is almost always fatal within the first few days of life). If the MSAFP screening test is positive, targeted ultrasound imaging of the fetus, and possibly amniocentesis, would be recommended for further evaluation. If they decided to have an amniocentesis, MSAFP screening would not be necessary because alpha-fetoprotein testing would be done on the amniotic fluid sample.

Finally, Sherman told the couple about a newer test called "non-invasive prenatal screening" (NIPS), which is based on analyzing fetal DNA found in the mother's blood. The test is highly accurate (greater than 99 percent) for detecting Down syndrome and a few other chromosome abnormalities, but it would not provide information about any other fetal disorders. Although highly accurate, NIPS is still considered a screening test, so if the results indicated a fetal problem, a confirmatory CVS or amniocentesis would be recommended. Sherman told the couple that the interpretation of NIPS was less complicated than maternal serum screening and would be less likely to indicate the need for a CVS or amniocentesis. He therefore suggested that the couple consider NIPS as their first choice. The first-trimester ultrasound and a second-trimester MSAFP blood test would still be recommended.

Sherman concluded this part of the discussion by telling Vanessa and Steve that a couple's decision whether to undergo fetal screening and what tests to have is based on many factors, including their tolerance for uncertainty about the health of the fetus, the risk of pregnancy loss from an invasive procedure, and the consequences of having an affected child. He told them that if they wanted, they could go directly to having either a CVS or amniocentesis as long as they understood and accepted the risks.

Based on their ethnicity, Sherman also recommended additional screening for the couple. Because Vanessa was of Greek descent, he suggested a complete blood count and hemoglobin electrophoresis to determine if she was a carrier for beta-thalassemia, a severe form of anemia (Steve would also be tested for this if she was positive). Steve was of Ashkenazi Jewish descent; because of his ancestry, Sherman recommended having carrier testing for cystic fibrosis, Tay-Sachs disease, Canavan disease, and familial dysautonomia. If Steve was a carrier for any of these diseases, Vanessa would also be screened. For couples in which both partners are carriers for any of these genetic conditions, the fetus has a one in four risk of being affected, and prenatal diagnosis for the disorder is available.

Sherman told the couple what he thought the best course of action was for them, and we believe that physicians should be prepared to do this or to give a direct answer to the question "Doctor, what would you do in this situation?" The response "I'm not you, so that's not relevant" is not useful. This is not like asking a waiter to suggest which item on the menu he or she likes best: taste in food varies widely and is not based on expertise. Patients, unlike diners dealing with waiters, seek out physicians because they are professionals and have special expertise, and it is reasonable to expect them to share their opinions about what they think is the best approach in a particular pregnancy. Of course, patients are not required to agree with their physician or to take their advice, but you will likely find your physician's advice useful in making your own decision.

Sherman asked Vanessa and Steve to consider all the information they had discussed, but not for too long. They would have to decide what course they wanted to take within the next few days if they wanted to maximize their options, especially first-trimester screening or CVS. Sherman told them they should call him if they had any questions, and he was pretty sure they would. He also provided written information that outlined everything he had discussed with them. Perhaps most importantly, he told them that there was no right or wrong answer; their decision should be based on what is best for them.

For Vanessa and Steve, what had started out as a dream day was ending in a decision-making nightmare. When Vanessa and Steve first learned of their pregnancy, they were extremely happy and thought mostly about adding the baby to their family. They just assumed that they were going to have a healthy baby. Sherman knew that they left his office worried about whether their baby might have a serious genetic disorder or birth defect. Modern medicine had succeeded in making them much more anxious and uncertain about the health of their baby-to-be. Couples in the near future will be faced with much more genetic and genomic information about their fetus and, paradoxically, more uncertainty. George thinks Sherman did a good job of informing Vanessa and Steve of their choices, but he also thinks it is important to have another knowledgeable person, such as a genetic counselor, to consult.

Sharing genetic messages with couples is, we think, imperative. But the question of how to maximize the good that genomic medicine can do while minimizing the potential harm is one that recurs as more and more genomic information is added to the practice of medicine, and this dilemma is not unique to prenatal care. New methods will have to be developed to mitigate information overload and help physicians convey genomic messages to pregnant patients in a helpful and meaningful way. We make some suggestions later in this chapter.

Karyotypes and Chromosomal Microarrays Analysis

For the past half century, the standard cytogenetic procedure for diagnosing chromosomal abnormalities has been a microscopic analysis of a complete set of chromosomes (normally forty-six) prepared from a dividing cell (for instance, a white blood cell or amniotic fluid cell). A systematized display of pairs of chromosomes in order of decreasing size is referred to as a *karyotype*. If, for example, three copies of chromosome 21 are present, termed trisomy 21, it is diagnostic of Down syndrome (figure 6.1).

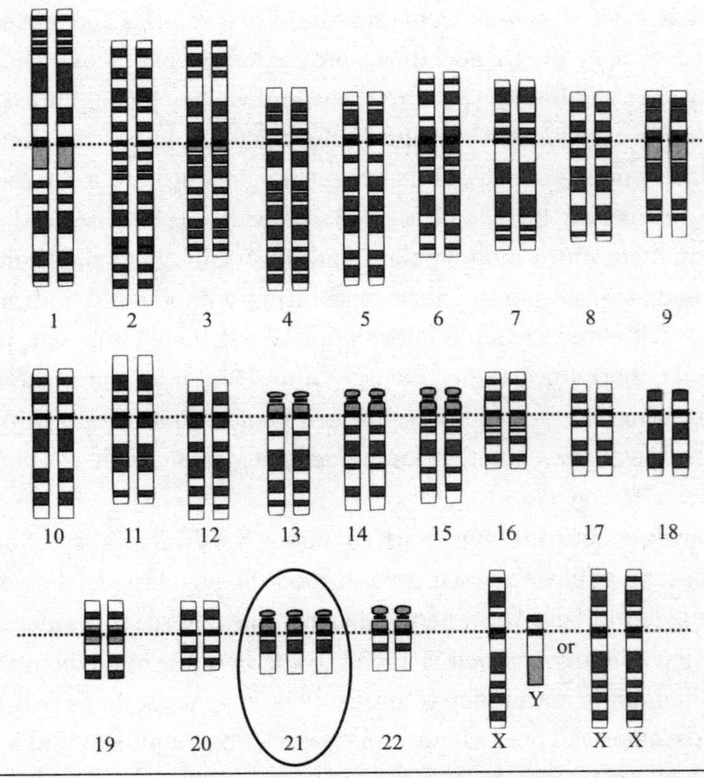

6.1 Karyotype for trisomy 21 or Down syndrome. Notice the three copies of chromosome 21. *Wikimedia Images*, June 23, 2006.

Chromosomal microarray analysis (CMA) technology detects not only entire extra or missing chromosomes but also small losses and gains of segments of DNA throughout the genome, while a karyotype analysis allows detection of extra or missing fragments of chromosomes, but only those large enough to be seen under a microscope. These losses and gains of segments of DNA are referred to as *copy number variants* (CNVs); CNVs can lead to genetic disorders associated with significant disabilities. CMAs are created by robots that print tiny dots arranged in a precise grid onto a glass slide. Each dot contains predetermined fragments of DNA sequences (probes) from known locations on each of the forty-six chromosomes. Microarrays typically contain millions of such dots. For prenatal diagnosis using CMAs, fetal DNA derived from amniotic fluid cells or a chorionic villi cell is "digested" (chopped up) with enzymes and then labeled with one color of fluorescent dye, while the control ("normal") digested DNA, derived from a person or pool of people with no known genetic abnormalities, is labeled with another color of fluorescent dye. Digested and control DNA is then added to the microarray probes. If the digested DNA is complementary to the microarray probes, the fluorescent dye lights up. The microarray is then placed in a special scanner that measures the relative brightness of each fluorescent dot (that is, the relative amount of fetal DNA versus control DNA). CNVs (duplications or deletions of fetal DNA) are detected as regions with higher or lower signal strength than the control sample. CMAs can also be "targeted," meaning that they are constructed to focus on CNVs of known pathogenicity instead of testing the entire genome.

The National Institute of Child Health and Human Development, a part of NIH, conducted a large study, called the MAStudy, to compare microarray analysis to karyotype analysis. A group of 4,406 women undergoing either CVS or amniocentesis was enrolled, and samples were split two ways: standard karyotyping was performed on one portion and microarray analysis on the other. In one in sixty cases where the karyotype was read as "normal," microarrays revealed a CNV considered of clinical importance. When prenatal diagnosis was prompted by a structural abnormality of the fetus seen on an ultrasound and the karyotype was interpreted as "normal," microarray analysis revealed a CNV in almost one in seventeen cases.

Clinical dilemmas are sometimes encountered when CMAs show "variants of unknown (or uncertain) significance," or VUS. A VUS is a change in DNA that has not yet been reliably characterized as benign or pathogenic. Sometimes the answer to the question "What is the chance of this VUS leading to a significant problem in my baby?" is that we just don't know. The vast majority of VUSes are likely to be just benign variants with no clinical consequences. On the other hand, even if a VUS is inherited from an apparently "normal" parent, the CNV may cause serious congenital and developmental abnormalities in the child. In such cases, finding a VUS could lead to questioning whether the parent who transmitted the VUS is really "normal," or if we need to search for health problems that have not been recognized. In the MAStudy, VUSes were found in 3.8 percent of all cases where the karyotype was read as "normal." A clinical advisory committee offered advice when VUSes were encountered. Sherman was a member of this committee. The case of Cindy and Mark (not their real names) provides an example.

Cindy, a healthy forty-year-old woman, underwent CVS during the twelfth week of her second pregnancy and participated in the MAStudy. Cindy and Mark's only child, a daughter, was reported to be normal. The family history revealed that Cindy had a first cousin with congenital deafness. On Mark's side, there were two first cousins, a brother and a sister, both reported to have autism and another first cousin with developmental delay and congenital heart disease. The chromosomal karyotype analysis from the chorionic villi indicated a normal female. However, CMA showed loss of a segment of DNA in one of the two chromosome 16s in the fetus. Cindy and Mark underwent CMA studies to see if one of them carried the loss of the same DNA segment, and it turned out that Cindy did.

The clinical advisory committee was consulted because similar losses of DNA segments on chromosome 16 have been found in both normal individuals and individuals with mental retardation, autism, seizures, and schizophrenia. An unaffected parent can have an affected child, even when both carry the same deletion. The committee decided that Cindy and Mark should be informed and counseled about the CMA findings. What the couple decided to do about the finding (that is, continuation or termination of the pregnancy) was not revealed to the committee.

Twenty-three other participants in the study who had received abnormal prenatal CMA results were interviewed by genetic counselor Barbara Bernhardt. Most reported being shocked, anxious, confused, and overwhelmed after receiving the news. One woman was quoted as saying, "You know, they're telling me there's something wrong, but they can't tell me what. . . . We wanted to know what that would mean for our son in the future. And they really couldn't tell us." Some women with uncertain results said they received conflicting information from counselors and physicians, and they also found contradictory information on the Internet. As one woman described it, "I started getting really panicky that the child I was carrying was going to be severely autistic with seizures and schizophrenia. . . . I would look online, and I met a geneticist and talked to an autism specialist. And frankly nobody could really tell me. . . . I ended up going to a crisis counselor because it was very stressful."

The limited time available for decision making makes it imperative to gather information quickly, including testing the parents to determine whether the fetal findings are inherited. If fetal abnormalities can be seen by ultrasonography, it adds suspicion that the VUS is deleterious, but lack of such findings does not guarantee the fetus is healthy. One woman explained, "I think what is so difficult about decisions—on top of the shock of it, the helplessness—is the timing. . . . We felt like we were desperately trying to build enough information to make an informed decision in a very fast amount of time, and that was very, very stressful for us."

Even when a CNV is known to be associated with a well-described genetic syndrome, making a decision about continuing the pregnancy may still be fraught with uncertainty. One woman, for example, terminated her pregnancy after her fetus was found to have a deletion of part of chromosome 22 (termed a 22q11.2 deletion). This deletion is associated with DiGeorge syndrome, a condition that may include heart defects, poor immune system function, a cleft palate, complications related to low levels of calcium in the blood, and behavioral disorders. Neither the woman nor the father carried the deletion. The woman said, "We still grapple with this because it is very much a spectrum of severity, very, very hard to predict what the outcome would be. . . . So that was very, very difficult for us because it made assessing our choices really hard." Half the women whose prenatal CMA results were uninterpretable or uncertain and who delivered normal-appearing infants had regrets about having the test in the first place—called "toxic knowledge." As one woman put it, "Since I had this uncertain microarray result. . . . If anything happens to him in the future . . . that will always pop up in my mind. . . . You just have to have a 'wait and see' attitude. . . . I'm a lot more vigilant."

Genomic information possesses a mythology of precision and determinativeness that it does not deserve. As the medical literature and databases associated with CNVs expand, such diagnostic uncertainties will be less frequent, but they will still occur. While studies continue, ACOG has recommended that couples choosing chromosomal microarrays should receive both pretest and posttest genetic counseling. Prenatal counseling is usually described as "nondirective" (that is, not for or against termination of the pregnancy), but this term has no meaning in the context of a genomic VUS. Physicians must share the uncertainty of diagnosis with their patients and advise them as best they can without guaranteeing a healthy baby.

Jay Katz has argued that sharing uncertainty is particularly difficult for physicians because they cling to the notions that uncertainty is grounded in ignorance and that it is their job to replace the patient's ignorance with their knowledge. Katz has suggested that physicians could learn from John Keats and his concept of "negative capability," the ability to live with uncertainties, mysteries, and doubts inherent in human life. Katz argues that physicians should be like poets in that the art of medicine is akin to discovering beauty in both its life-affirming and life-destroying dimensions; the science of medicine seeks to discover truth in the "beauty of discovery and in the ugliness of ignorance." Katz believes that dealing with uncertainty is the key to doctor-patient conversations and that medical education has neglected training in "how to truly cope with uncertainties" without "becoming paralyzed by them." This is necessary for both physicians and patients, Katz suggests, because "it is the legacy of science that scientific activity produces not only new knowledge but also new ignorance," and physicians need to acknowledge both to their patients. These lessons may prove particularly valuable when adopting new methods of prenatal diagnosis where, quoting Katz again, "the disregard of uncertainty defeats the sharing of the burden of decision" and promotes evasions and even lies that can make "meaningful disclosure and consent a charade."

For more than three decades, finding a noninvasive way to analyze the genetic makeup of the fetus by identifying fetal cells in the blood of the pregnant woman has been a scientific quest because it could permit prenatal diagnosis without risk to the fetus. Sherman's laboratory, as well as others, showed that cells from chromosomally abnormal fetuses, including trisomy 21 and trisomy 18, could be detected in maternal blood during pregnancy as early as ten weeks of gestation. An evaluation by the National Institute of Child Health and Human Development (NICHD) concluded that technological advances were needed before fetal cells in maternal blood could be used for routine clinical application. The quest to dependably collect fetal DNA from maternal blood has been taken up by private biotech companies, one of which Sherman worked with as scientific director. In 1997, it was first reported that fragments of cell-free fetal DNA (cfDNA) circulate in the maternal blood during pregnancy beginning in the early first trimester. About 10 percent of the DNA in maternal plasma is now known to be of placental origin. This offers the possibility of noninvasive prenatal screening by taking a blood sample from the pregnant woman.

The technology currently used for commercial testing is *massively parallel genomic sequencing*. This is a highly sensitive assay that rapidly quantifies millions of DNA fragments from maternal blood. Bioinformatician Titus Brown has likened this process to shredding one thousand copies of *A Tale of Two Cities* in a wood chipper and then putting them back together. ACOG considers cfDNA an option for primary screening in women at increased risk for having a child with a chromosomal disorder but does not currently recommended it for general obstetric patients. A limitation of cfDNA testing is that it provides information regarding only trisomy 21, 18, and 13, as well as the number of X and Y chromosomes, and this should be explained when counseling patients. More recent data has shown that cfDNA can be accurate in all pregnancies, not just high-risk pregnancies. Performing a cfDNA test should be an informed patient's choice after pretest counseling and can be considered mainstream obstetrical care. This technique is being so rapidly introduced into clinical practice that it has been reported that the rate of CVS and amniocentesis has dropped significantly in the past 2 years, perhaps as much as 50 percent.

On the other hand, at least some physicians and their patients do not understand how these screening tests work, and they have aborted healthy fetuses as a result of a screening test without following it up with a diagnostic test. A story published on the front page of the *Boston Globe*, for example, concluded that "companies selling the most popular of these screens do not make it clear enough to patients and doctors that the results of their tests are not reliable enough to make a diagnosis." The article features Stacie Chapman, who was told by her physician, Jayme Sloan, that the test had determined that her fetus had trisomy 18, a condition incompatible with life, and that the screening test she had, Sequenom's MaterniT21 Plus, had a 99 percent detection rate. What neither she nor her physician appreciated was that there was a good chance the test was wrong. After discussing trisomy 18 with her husband, Stacie made the decision to terminate her pregnancy. Before she did, however, her physician called her back and urged her to have a diagnostic test, which showed that her future son did not have trisomy 18 after all. In Stacie's words, "He is so perfect. . . . I almost terminated him."

Other women were not so lucky and did terminate healthy fetuses on the basis of a screening test alone. Michael Greene, director of obstetrics at Massachusetts General Hospital, observed that the marketing of tests has created problems: "The companies have done a very poor job of education . . . failing to make clear that it is screening testing with very good but inevitably not perfect test performance . . . and that doctors are recommending, offering, ordering a test they do not fully understand." Not only are some women aborting healthy fetuses, but others are giving birth to affected fetuses even after having the screening test. Belinda Boydston, for example, says she gave birth to a child with trisomy 18 even after being assured that the fetus had only a 0.01 chance of having the condition. One solution to dealing with these unexpected results is not to use these screening tests without prior genetic counseling.

The Future of Prenatal Diagnosis

Within the next few years, noninvasive prenatal detection for the majority of recognized genetic disorders will likely become a reality. The fact that the entire fetal genome is represented in maternal blood opens up the possibility of obtaining fetal DNA fragments that can be assembled into a complete fetal genome readout. A number of substantial hurdles, including the lack of genetic counseling, remain before noninvasive fetal whole-genome sequencing is introduced into clinical practice. First, the cost will have to drop substantially. In this regard, targeting selected genomic regions may prove more efficient and cost effective than deriving the whole genome. On the other hand, isolating and analyzing fetal cells from maternal blood would be more straightforward and presumably less costly. It is likely that noninvasive whole-genome sequencing of the fetus will eventually be done at a cost of less than $1,000, and sequencing the entire genome may be the most efficient approach. We are already at the point where we can identify "3,600 genes for rare Mendelian disorders, 4,000 genetic loci related to common diseases, and several hundred genes that drive cancer."

A critical consideration for prenatal genetic diagnosis, whether by invasive or noninvasive means, is deciding which conditions warrant testing. There is little doubt that noninvasive prenatal screening will be viewed by both the public and physicians as a positive advance primarily because it avoids the risk of miscarriage produced by invasive tests. However, philosopher Evelyne Shuster has suggested that whole-genome fetal testing could act as a metaphorical "toll booth" on the road to childbirth, and that as testing for hundreds or even thousands of genetic conditions becomes feasible, it will become more and more difficult to decide if any fetus should be considered "healthy" enough to warrant carrying to term. Are we entering a new phase of eugenics—private rather than government-sponsored—with the question of who deserves to be born decided in private doctors' offices? Is private eugenics the price we pay for private decisions? Physician-philosopher Georges Canguilhem has suggested that genomics always begins with a "dream . . . to spare innocent[s] . . . the atrocious burden of producing errors of life. At the end there are the gene police, clad in the geneticists' science." To the extent he is correct about government's natural tendencies, it is our task to recognize and oppose the tendency toward a new eugenics.

Will noninvasive prenatal testing become "normalized" and routine because of its ease and safety? We think it is inevitable that reasonably priced and accurate genetic and genomic tests based on fetal cells or cfDNA extracted from maternal blood will be offered to all pregnant women. The primary motivation, however, will likely not be medical practice standards but fear of medical malpractice lawsuits. Obstetricians will likely fear a malpractice lawsuit if the fetus is born with a genetic abnormality for which the couple can credibly explain to a jury that they would have had the genetic test (had they known about it) and terminated the pregnancy. We both believe that physicians' fear of the extremely unlikely prospect of such a lawsuit is a terrible way to set medical practice standards. Rather, we support professional organizations that set screening standards for their members—and whose members follow them.

Not all screening and testing decisions, even for fetuses, will be made in the context of the doctor-patient relationship. Aggressive marketing and consumer demand will likely drive the utilization and adoption of new genomic technologies. We are already seeing cfDNA noninvasive prenatal screening companies reaching out to pregnant "consumers" on YouTube, Facebook, and Twitter. Companies that can consistently predict fetal sex early in pregnancy seem likely to find a consumer market in the United States. A market is also likely to exist for noninvasive prenatal paternity testing. Some women may decide that without paternity testing they would terminate the pregnancy because they could not be sure of the identity of the "right father" for the fetus, a story line at the center of the second season of the popular zombie series, *The Walking Dead*.

Translating and conveying expanding genomic information into useful knowledge is becoming increasingly challenging. Taking a family history and, when indicated, recommending tests to identify carriers of genetic diseases are now standard in obstetric care. Today's carrier screening tests usually focus on conditions that occur in the ethnic groups of one or both prospective parents. With noninvasive prenatal testing, we foresee a time when, rather than screening parents to identify carrier status for genetic disorders, fetuses will be directly tested to determine whether they are affected. (This would allow fetal testing without determining paternity.) It will then become increasingly difficult—if not impossible—to inform those offered screening or testing for reproductive purposes about all the genetic and genomic information that can be obtained and the implications of that information.

Our current model for prenatal screening and diagnostic testing requires pretest counseling prior to obtaining informed consent, and the obligation to counsel can be seen as inherent in the fiduciary nature of the doctor-patient relationship. For ordinary medical procedures, the physical risks and treatment alternatives (those things that might lead a patient to reject therapy or choose an alternative) are the primary items of information that must be disclosed and should be discussed. Self-determination and rational decision making are the central values protected by informed consent. In the setting of reproductive genetics, what is at stake is the right to decide whether or not to have genomic testing, with emphasis on the right to refuse if the potential harm (in terms of stigma or unacceptable choices, including abortion) outweighs, for the individual or family, the potential benefit.

Hundreds of new genomic screening and diagnostic tests, including exome and whole-genome sequencing, addressed in the following chapter, will compete for introduction into routine clinical practice. As we outlined at the beginning of the chapter, critical questions include these: What information should be provided to which patients? When should it be provided? How and by whom should it be conveyed? It will soon be impossible to do meaningful counseling about all available genomic testing. Giving too much information (information overload) can amount to misinformation and make the entire counseling process misleading or meaningless.

To prevent disclosure from being pointless or counterproductive, we believe that information-sharing strategies based on general or "generic" consent should be developed for genetic and genomic screening and diagnostic testing. Their aim would be to provide sufficient information to permit patients to make informed decisions yet avoid the information overload that could lead to "misinformed consent."

Traditionally, the goals of reproductive genetic counseling are to help the person or family to

- *comprehend the medical facts*, including the diagnosis, the probable course of the disorder, and the available management choices;
- *appreciate the way heredity contributes to the disorder* and the risk of recurrence in specified relatives;
- *understand the options* for dealing with the risk of recurrence;
- *choose the course of action* that is appropriate in view of their risk and their family *goals,* and act in accordance with that decision; and
- *make the best possible adjustment* to the disorder in an affected family member or to the risk of recurrence of the disorder.

As we saw with Vanessa and Steve, even knowledgeable couples can become confused, frustrated, and anxious if faced with multiple options for genetic screening and testing. An approach based on "generic consent," an approach we described in the *New England Journal of Medicine* twenty years ago and still believe in, would not even attempt to describe each of hundreds or thousands of genetic conditions and anomalies to be screened and tested for, but would instead emphasize broader concepts and common-denominator issues in genetic and genomic screening.

We envision a doctor-patient relationship in which patients are told of the availability of a panel of genetic and genomic tests that can be performed on a single blood sample, either for carrier screening or noninvasive prenatal testing. Couples would be told that these tests focus on disorders that involve serious physical abnormalities, mental disabilities, or both. Several common examples would be given to indicate the frequency and spectrum of severity of each type or category of genetic condition for which screening or diagnostic testing was offered. Conditions such as spina bifida and cleft lip, chromosome abnormalities such as Down syndrome and trisomy 18, and single-gene disorders such as cystic fibrosis and Tay-Sachs disease, might be chosen as representative examples.

In the course of counseling, important factors common to all prenatal screening and diagnostic tests would be highlighted. Among these are their limitations, especially the fact that negative results cannot guarantee a healthy infant. For screening tests, the couple needs to know that additional, invasive tests may be needed to establish a diagnosis or clarify confusing or uncertain results. Other considerations that need to be discussed are options (such as adoption, egg or sperm donation, abortion, or acceptance of risks); the costs of testing; and issues of confidentiality, including potential disclosure of the results to other family members. If the testing is for carrier status, and a recessive gene is detected in the woman, it must be emphasized that her partner should also be screened (and this could bring up the issue of nonpaternity).

For prenatal diagnosis, couples should understand that abortion of an abnormal fetus is available but not the only option. For couples for whom abortion is unacceptable for any indication, prenatal diagnosis may still provide important information. For example, if it is known that a baby will be delivered with serious birth defects, choosing delivery at a tertiary medical center that offers specialized care may optimize the infant's outcome. In some cases, knowing that the fetus has a very serious and incurable disorder may alter obstetrical care for the mother. For example, if the fetus has trisomy 18, there is a high likelihood that fetal heart monitoring during labor would show an abnormal tracing. Knowing that the fetus has trisomy 18 could avert performing a cesarean delivery because it would not benefit the infant.

Generic consent to genomic screening and diagnostic testing can be compared with obtaining consent to perform a routine physical examination. Patients know that the purpose of the examination is to locate potential problems (the doctor is looking for trouble, and the patient is hoping that no trouble will be found) that are likely to require additional follow-up and that could present them with choices they would rather not have to make. The patient is not generally told, however, about all the possible abnormalities that can be detected by a routine physical examination or routine blood work, only the general purpose of each. On the other hand, tests that may produce especially sensitive and stigmatizing information, such as screening blood for the human immunodeficiency virus (HIV), should not be performed without specific consent. Similarly, because of its reproductive implications, genetic testing has not traditionally been carried out without specific consent.

What is central in generic consent for genomic screening and diagnostic testing is not a waiver of the individual patient's right to information. Rather, it would reflect the physician's fiduciary obligations to the patient and a conclusion that the most reasonable way to conduct genomic screening and diagnostic testing for multiple diseases is to provide basic, general information to obtain consent for the screening, and much more detailed information on specific conditions if they are detected. Since, in the vast majority of cases, no such conditions will in fact be found, this method is also the most efficient and cost effective.

Some patients may want more specific and in-depth information on which to base their decision regarding testing. It is therefore essential to build into the consent process ample opportunity for patients to obtain all the additional information they want to help them make decisions. Clinicians, of course, must be open and responsive to the concerns and questions of patients. Counseling could be provided in person by a physician or other qualified health professional. Alternatively, Internet-based audiovisual aids could be used, which could help ensure consistency in the information provided, improve efficiency, and respond to the shortage of genetic counselors.

Generic consent for genetic and genomic screening and diagnostic testing should help prevent overloading patients with information and wasting time on useless information, especially for carrier screening. It would not, however, solve what is likely to be an even more central problem in prenatal genetic testing: are there genetic conditions for which testing should not be offered to prospective parents? Examples might include genes that predispose a person to a particular disease that will not appear until late in life (such as Alzheimer disease, Parkinson disease, or breast cancer). From the perspective of the fetus, life with the possibility—or even a high probability—of developing these diseases in late adulthood is much to be preferred to no life at all. Thus, in this case, unlike that of the fetus with trisomy 18, for example, no reasonable argument could be made that precluding abortion by denying this information could amount to forcing a "wrongful life" on the child.

We should, nonetheless, directly and publicly address the question of whether there are conditions for which screening prospective parents or testing fetuses should not be offered as a matter of good medical practice and public policy, regardless of the technical ability to do such testing and the wishes of the couple. Offering genetic and genomic screening and diagnostic testing to assist couples in making reproductive decisions is not a neutral activity, but rather implies that some action should be taken on the basis of the results of the test. Simply offering carrier screening for breast cancer or colon cancer genes in the context of preconception care, for example, suggests to couples that artificial insemination, adoption, and even abortion are all reasonable choices if they are found to be carriers of such genes. On the other hand, because of a personal experience with a family member who suffered from one of these adult-onset diseases, a particular couple might see abortion as a reasonable choice under such circumstances. A practice of keeping such information away from all couples would not be justified. However, in general we do not believe that pregnancies in women who want to have a child should be terminated for adult-onset diseases. We are all going to die of something, and if we live long enough, that something will have a major genetic component. There can be no perfect genome, and the search for it in a fetus will inevitably fail.

A standard of care for genomic screening and diagnostic testing, as well as a standard for informed consent in the face of hundreds or thousands of available genomic tests, will inevitably be set. We believe the medical profession should take the lead in setting such standards and that, with significant public input (not common today) and support, the model of generic consent for genomic screening and diagnostic testing will ultimately be accepted.

After you finish this chapter, it may seem that there is no such thing as a "healthy" fetus. The practice of prenatal screening and diagnosis is founded on looking for problems. However, with the rapidly increasing number of genomic and other types of testing during pregnancy, it is increasingly likely that one or more "abnormal" results will turn up, at least some of which may be "serious." Given all this anxiety in both perspective parents and their physicians, it may seem remarkable that the vast majority of babies are just fine. Perhaps the hardest lesson for us to accept is that the future is uncertain. We all want reassurances that our future, especially the future of our loved ones, including our future children, will be all right.

The future is unknowable, and there really are no guarantees, no matter how much testing is done. This leads to stress and anxiety, and we must live with ambiguity. This is the price we pay for the information we receive from testing and the opportunities it gives to make potentially better choices for our lives. Keats can help us here, as he helped Katz with his idea of negative capability. In the most anthologized poem in the English language, *To Autumn*, Keats accepts mortality but finds beauty and truth in aging: "Where are the songs of Spring? Ay, where are they? Think not of them, thou hast thy music too." We are in the early spring of prenatal genetic screening, and if done right, future generations will look back on these times in genomics as the "songs of Spring" that opened a conversation we will think of with fondness and pride.

In the next chapter, we present even more genomic screening tools that are used to look for specific problems—but this time problems that are treatable if detected in a child or newborn. In the context of newborn screening, we again address the question of whether and how expanding the number of genomic screening tests can reasonably be introduced into medical and public health practice. As you have probably already guessed, although we can divide genomic screening by populations, including adults, children, newborns, and fetuses (and even preimplantation embryos, as discussed in the previous chapter), once genomic screening is viewed as reasonable for any of these populations, it is almost inevitable that at least some physicians and patients will uncritically see it as reasonable for all of them.

WHEN THINKING ABOUT PRENATAL GENOMIC SCREENING, CONSIDER THESE THOUGHTS

Prenatal screening has relied primarily on measurement of proteins in the pregnant woman's blood but is fast moving into cell-free placental DNA in the pregnant woman's blood.

We may soon be able to use cell-free DNA from maternal blood to do noninvasive whole-genome sequencing of the fetus.

There is no perfect genome and no genetically perfect fetus.

Informed consent is fostered by physicians sharing uncertainty with patients.

New consent models for prenatal screening will have to be developed and should be judged on whether they improve the patient's right to decide.

CHAPTER 7

Genomic Messages from Newborns and Children

There is no difference between men, in intelligence or race, so profound as the difference between the sick and the well.

—F. Scott Fitzgerald, *The Great Gatsby* (1925)

There are two primary ways to use genomics for our children. The first is to attempt to diagnose a child whose illness cannot be diagnosed by conventional medical means. There are a handful of famous cases in which children who were very sick with unidentifiable, rare conditions were successfully diagnosed and treated using whole-genome sequencing, and more are continually being added to the list. The second, is to use genomic sequencing at birth or during childhood to screen healthy children for treatable conditions. In this chapter, we examine both these uses of genomics. In the first we will examine a classic challenge in medicine: using our vastly increased genomic information to diagnose a rare childhood disorder. In the second, newborn screening, we address what to screen newborns for, including the complex role of the state in mandating screening and the role of the parents in providing informed consent to genomic screening. As we will see, because newborn genomic screening can detect genetic risks that will not be relevant until adulthood, children have a dignity interest in making their own decisions about learning this information when they become adults. Likewise, because the genetic information of the newborn also contains genetic information about the parents, the parents have a personal health interest beyond the interest they have in their child's health.

Next-generation sequencing encompasses a variety of technologies that enable rapid sequencing of many large segments of DNA, up to and including entire genomes. A distinction must be made between *whole-exome sequencing* (WES) and *whole-genome sequencing* (WGS). WES is sequencing the "exome," that portion of the genome that codes for proteins. The exome makes up only about 1 percent of the genome, but it is the part most likely to include mutations that increase the risk of diseases. WGS implies sequencing most or all of the DNA content (protein-coding as well as the remainder of the genome), although there may be components of the genome that are not included in a present-day "whole-genome sequence." Costs increase as one goes from WES to WGS, though the difference will decrease over time. We use the term *genome sequencing* to include both WES and WGS. As the cost of sequencing goes down and the number of potentially useful discrete genetic tests increases, a tipping point will be reached at which obtaining data by genomic sequencing will be more efficient than looking for mutations in individual genes. This change will be reflected in all applications of genomic sequencing in clinical medicine, including newborn screening. Of course, costs cannot control use: if a test has no practical clinical use or causes more problems than it solves, it should not be used even if it is free.

Genome Sequencing of Ill Children

Genome sequencing is not reasonable for healthy children but may be appropriate for a particular ill infant or child. The ACMG states that

WES/WGS should be considered if a patient is believed to have a likely genetic disorder, but specific genetic tests available for the clinical picture have failed to arrive at a diagnosis. The ACMG and American Academy of Pediatrics (AAP) recommend that parents be informed of potential benefits and potential harms, and their consent obtained. Of course, whether these groups recommend this or not, informed consent is a legal and ethical requirement. The benefits include determining possible preventive or therapeutic interventions, decisions about surveillance, clarification of the diagnosis and prognosis, and recurrence risks. It is also acceptable to perform pharmacogenetic testing to help with drug selection, appropriate dosing, and other safety and efficacy issues. Potential harms can occur if the results lead to pursuing unproven treatments, particularly if they are ineffective or have significant adverse effects. These recommendations properly focus on the best interests of newborns and children.

Rapid genome sequencing (in about forty-eight hours) for disease gene sleuthing has been used to provide essential information in the newborn intensive care unit (NICU). For acutely ill babies, coming up with a quick diagnosis of a genetic disorder can be vital. There are about five hundred diseases involving mutations in single genes for which there are therapies. The proper treatment can usually be determined without resorting to genomic sequencing; nonetheless, genomic sequencing can help when more conventional diagnostic tests fail to diagnose a potentially treatable condition. Unfortunately, most cases in which sequencing has been used in the NICU for rapid diagnosis have to date involved genetic disorders for which there is no effective treatment. For example, an infant girl developed seizures within an hour of being born. Despite intensive efforts to locate a possible infection and tests looking for metabolic diseases and chromosomal abnormalities, all results came up negative. A brain MRI was normal, but an electroencephalogram (a brain wave test) confirmed abnormal brain activity. Multiple antiseizure drugs were tried, to no avail. The infant had to be put on a ventilator because of low oxygen levels and a slow heart rate. At this point, whole-genome sequencing was done to attempt a diagnosis. A mutation was found in both copies of a gene on chromosome 7 designated as *BRAT1*. Both parents were tested and found to be carriers for the same mutation. After lengthy discussion, the parents requested the withdrawal of life support.

In cases like this one, rapid diagnosis by genomic sequencing of an untreatable and inevitably fatal disorder allows physicians and the family to stop looking for a cause. The decision can be made to move the infant to a more appropriate setting, where the parents can be given an opportunity say good-bye. Sometimes discovering that the parents are carriers for the same mutation indicates that they are at risk for having another affected child, which can be important for future reproductive decisions. Current research indicates that, using clinical genome- and exome-sequencing techniques, a causative genetic variant can be identified in approximately 25 percent of pediatric patients who have been identified as having a condition that is likely due to a single-gene genetic disorder. Because finding the genetic cause is relatively unusual and other conditions may be identified (so-called incidental or secondary findings), it is critical that realistic pretest counseling be conducted with the parents. Most importantly, even when the causative genetic variation is discovered, it is still unusual to have the diagnosis lead to a change in medical management or an improvement in prognosis. Treatment, of course, is the ultimate goal—but that goal is, for now at least, still in the distant future for most children. Two studies have been able to make a molecular diagnosis using exome sequencing in about 25 percent of sick children, and a medically actionable finding in about 5 percent of the children. Commenting on one of the studies, clinical geneticist Jonathan Berg noted that "the difficulty of interpreting the clinical significance of genomic variants" remains a challenge, and "there is much to learn before [genomic sequencing] can be applied more universally."

Nonetheless, there have been spectacular treatment successes. Retta Beery, the mother of twins, Alexis and Noah, described her diagnostic odyssey to the President's Bioethics Commission. Her twins were born in 1996, and it soon became clear from their jerky movements that something was seriously wrong. Over the following two years, she and her husband, Joe, took the twins to a variety of specialists, and they endured countless tests and surgeries with little result. In 2002, with a diagnosis of Segawa's dystonia (DOPA-responsive dystonia) the doctors began a course of treatment to increase the childrens' brain dopamine, which dramatically improved their health. In 2009, Alexis developed severe breathing problems, and in 2010 the Beerys went to Baylor University Medical Center for whole-genome sequencing. The sequencing revealed a rare and only recently recognized genetic cause of dystonia. This information enabled their neurologist to successfully treat the twins with an over-the-counter supplement.

Another highly publicized success story involves Nicholas Volker. When Nicholas was fifteen months old, his parents sought care for him because of poor weight gain, inflammation, and a large draining abscess around his anus. Antibiotics did not help, and his condition deteriorated. At two and a half years, Nicholas was sent to a major medical center, where he was found to have severe growth stunting and malnutrition. The presumed diagnosis was a severe form of Crohn's disease, an inflammatory disease that usually affects the intestines. The cause of Crohn's disease is not well understood, and there is no definitive diagnostic test; it is thought to result from interactions among genetic, environmental, immunological, and bacterial factors. Despite treatment, his condition continued to deteriorate. A colostomy, a surgical procedure that brings one end of the large intestine out through the abdominal wall to divert fecal material, was performed. He developed sepsis (a potentially fatal response to bacteria or other germs) and required four weeks in the intensive care unit. At age four, he was again admitted to the hospital, and his entire colon had to be removed.

Over the next several years, Nicholas continued to require hospitalizations and experienced severe, potentially lethal complications. To keep him alive, a highly aggressive and risky approach would need to be tried: a bone marrow transplant using donor cells. The success of this approach would depend on knowing the exact underlying cause of the child's disease.

Whole-exome sequencing was performed on his DNA, identifying over 16,000 variants compared to the human genome reference. Using highly sophisticated analysis, a single substitution of a G (guanine) for an A (adenine) was found in the *XIAP* (X-linked inhibitor of apoptosis) gene, found on the X chromosome. The XIAP protein encoded by the *XIAP* gene plays a central role in the inflammation process. Further testing showed that Nicholas had a deficiency of XIAP protein that was causing his immunodeficiency disease. Based on these findings, at five years eight months, Nicholas underwent a bone marrow transplant using donor cells. Within a few months he was on a normal diet and had no recurrence of his bowel disease.

Together with the Beery twins, this case shows how powerful genomic sequencing information can be in diagnosing and successfully treating a rare disease—an example of personalized genomic medicine at its best. In the case of Nicholas, it was life saving. However, in the 2014 review article, these two cases were the only ones the authors could identify in which whole-genome screening of children led to a new treatment regime that dramatically changed the child's clinical outcome. Of course there will likely be many more; our only point is that we are in very early days of using genomics to diagnose and treat children with rare genetically-determined diseases.

What if your child is healthy? Does it make any sense to screen him or her for "good genes" that might predict special talents? The Atlas First SportGene® Test presents this question. We discuss it in some detail because it can stand in for almost any special talent, including music, mathematics, or ballet. The company claims that the test "is geared specifically to show athletes, trainers and interested individuals where their genetic advantage lies." The SportGene® Test focuses on a gene called *ACTN3,* which is involved in forceful contraction of fast-twitch muscle fibers used during sprint-type activity. In a recent meta-analysis, which combines the results of a number of studies that address a set of related research hypotheses, a variant of the *ACTN3* gene called the RR genotype was found to be more common among sprint and power athletes compared to controls. Can we therefore conclude that a child found to have the RR genotype of the *ACTN3* gene will be a more gifted soccer player? The answer is no. This is because an "athletic champion" is the result of multiple factors, including the combined influence of hundreds of genes that are expressed in many organs, such as skeletal muscles (fiber characteristics), lungs and blood (oxygen diffusion capacity), connective tissue (tendon stiffness), heart (maximal pump capacity), thyroid (metabolic rate), pancreas (insulin secretion and glucose control), adrenal glands (adrenalin surges), and brain and nerves (balance, reflexes, and pain tolerance), among others. Each of these individual genes may have multiple DNA variants that can result in positive or negative effects on the overall level of athletic performance. Any given gene variant may interact differently with other gene variants, so-called gene-gene interactions. In addition, environmental influences (for example, sports culture, nutrition, training intensity, coaching and instruction, temperature, and exercise equipment, to name just a few) are critical, as is just plain luck.

Variation in any single gene would likely have negligible effect on sports performance. For example, it is estimated that the *ACTN3* gene accounts for only about 2 percent of variance in muscle performance. Even complete deficiency of the *ACTN3* gene product, α-actin-3 protein, does not appear to preclude elite performance, as demonstrated by a Spanish Olympic long jumper with both copies of his *ACTN3* gene having the X ("null") variant. As Carl Foster, director of the Human Performance Laboratory at the University of Wisconsin – LaCrosse and coauthor of several *ACTN3* studies, has said, "If you want to know if your kid is going to be fast, the best genetic test right now is a stop watch. Take him to the playground and have him race the other kids." There will undoubtedly be companies trying to sell you tests to determine your child's intellectual ability or musical talent. Like sports ability, there are much better tests for intelligence and musical ability: each involves directly testing the talent by actual performance.

Before moving from child to newborn screening, we should say a few words about incidental or secondary findings, genetic findings that are not what you are looking for but are discovered when doing exome or genetic sequencing (like that done for Nicholas Volker). Because screening children so rarely yields results that directly lead to more effective treatment, it has been suggested that whenever this type of screening is done, the testing lab should also look for other genes that might cause trouble to the child in the future, or for which the child's parents might be at risk. We think this strategy is legitimate provided the parents get genetic counseling describing the nature of the added search and have the basic right of informed consent, including the right to refuse all or parts of the secondary testing. The original 2013 recommendations of the ACMG disregarded basic legal and ethical rights of informed consent, asserting that fifty-six specific genes should routinely be analyzed for mutations whenever genomic sequencing was performed on a patient. These included thirty-one genes associated with cardiovascular risks, twenty-three mutations associated with cancers and noncancerous tumors, and two genes associated with malignant hyperthermia (a muscle disorder that results in high fever and severe muscle contractions when general anesthesia is given). We strongly disagreed that a medical group could or should attempt to compromise the legal rights of their patients. We were pleased when, a year later, the college amended its recommendations to require informed consent complete with an opt-out provision for the supplementary hunt for the fifty-six genes. The new genetics, as powerful as it is likely to be in the future, is not so powerful or exceptional that it can or should change basic patient rights, including the right not to have medical tests performed without informed consent. As we put it in *Science* with our colleague Susan Wolf, a law professor: "Starting down the path of unconsented testing and reporting in clinical genomics leads to grave difficulties and should not be done without more careful analysis. . . . The era of medical genomics requires a trusting partnership with patients, based on respect for their rights." We could have added that informed consent is a matter of law, not a matter to be determined by physicians, no matter how distinguished or knowledgeable they might be about genomics.

Genome sequencing of children will nonetheless likely be an easy sell to many parents, especially if it is marketed as the "Rosetta Stone" that can help ensure the best health for their children. It seems that whenever new technologies become available, we are anxious to adopt them. Recall that two fundamental characteristics of American medicine are that it is technologically-driven and individualistic. Even if we don't know how we will use it, having access to more technology (and to more information) almost always seems to be a good thing, especially when it relates to our health. Nonetheless, unless your child has a serious condition that cannot be diagnosed without genomic screening, the right not to know is today much more important than the right to know. Almost all of the information genomics can now provide to parents of healthy children is either useless or potentially harmful. Using sequencing technologies, it is now known that every individual has 3–4 million variants. Virtually all of these variations are meaningless today. Diverting the attention of health care providers and parents to do exhaustive analysis and reporting of data for which we have no understanding, and which contains no known medical relevance, is unwarranted. There are also potential harms.

Some parents may view their child as "abnormal," even though there is no evidence of a problem. In adults we call this the "worried well"; in the case of newborn screening, they could be called "worried parents of the well." In this regard we agree with the conclusions of H. Gilbert Welch in his profound book *Overdiagnosed: Making People Sick in the Pursuit of Health*. Welch noted that although most of us have normal phenotypes, "each of us can be shown to be at high risk for some disease. So the new world of personal genetic testing has the potential to make all of us sick." This is a significant problem, especially when it involves labeling children as sick. That is because Fitzgerald's observation at the beginning of the chapter is correct: there really is no more profound difference in people than the difference between the sick and the well—and the last thing we should want to do is make ourselves, and our children, think we are sick when we're not. As Welch provocatively concludes, "Ironically, the healthiest populations may be those that know nothing about their DNA." The problem of variations of uncertain significance (VUS) is not unique to genetic testing, and there is inevitably a learning curve whenever we introduce new diagnostic tests into medical practice. For example, in magnetic resonance imaging, variants in healthy body structure are frequently encountered. Usually we do not waste time documenting such variants pixel by pixel but instead focus on those we currently understand to have clinical significance. The same should go for a VUS in the genome. Geneticist William Gregory Feero puts it another way, arguing (like Donald Rumsfeld on the weapons of mass destruction in Iraq) that there are many "known unknowns" and "unknown unknowns" that we will have to learn about before whole-genome screening can be integrated into clinical practice. The question for now is whether WGS data "will decrease uncertainty and improve outcomes or merely exponentially increase the complexity of clinical care."

Newborn Screening

Produced almost two decades ago, the movie *Gattaca* still captures the uneasiness we all feel about genes in terms of destiny, discrimination, and eugenics. In *Gattaca,* parents using IVF and preimplantation manipulation can choose at least some of the genetic makeup of their children. To ensure that a "valid" offspring is produced, embryos are weeded out in the laboratory if they are found to have a "critical predisposition to any major inheritable disease." There are "potentially prejudicial conditions," such as premature baldness, myopia, alcoholism, addictive susceptibility, and propensity for violence or obesity, that can be modified at the embryo level. Those who are conceived naturally are considered "invalid." In *Gattaca,* they are described as having "discrimination down to a science."

Vincent Freeman is "invalid," his parents having "put their faith in God's hands rather than those of the local geneticist." Immediately after his birth, a nurse pricks his heel. A drop of blood is inserted into an analyzing machine that prints out the baby's health future. Antonio, his father, asks, "What's wrong?" Vincent's voice-over continues: "Of course, there was nothing wrong with me. Not so long ago I would have been considered a perfectly healthy, normal baby. Ten fingers, ten toes. That was all that used to matter. . . . [Now] only minutes old, the date and cause of my death was already known." Vincent's newborn test results are read out loud by the nurse in the delivery room: "Manic depression, 42% probability; attention deficit disorder, 89% probability; heart disorder, 99% probability; life expectancy, 33 years."

Is *Gattaca*-like newborn screening in our future? Should it be? We begin with a true story about a newborn named Kimberly. Christina and Robert couldn't have been happier. The nurse had just come into Christina's hospital room with their new daughter, Kimberly, who had been taken to the nursery for her discharge examination. Christina noticed that Kimberly had a small adhesive bandage on her right heel and asked the nurse about it. The nurse explained that a few drops of blood had been taken from Kimberly for newborn screening for a number of rare diseases, and that it was done because "it's the law."

We have now left the realm of medical practice and entered into the realm of public health—where decisions are made by state governments to protect populations rather than individuals. In the context of genetics, we seem to be in the process of abandoning personalized medicine before it has even been introduced into the clinic, skipping directly to population health. This is certainly true of newborn screening, even though routine screening of newborns has had little impact on health at the population level. There are three things Christina and Robert (and all new parents) should know for sure. First, thousands of babies in the United States (out of millions) will have a very rare condition that could be detected (and hopefully treated) by newborn screening. Second, the chance of any individual baby having one of them is vanishingly small. Third, new screening tests always bring with them new worries, including those spawned by false-positive results (detecting a disease where none actually exists). See Appendix B for more on screening tests.

Mandatory public health screening laws are state laws because the states retained this arena of authority (sometimes called the "police powers") when they delegated government power to the federal government through the Constitution. That means there can be, and is, variability among the states on both the content of newborn screening and parental consent to it. State-mandated newborn screening is performed on more than 4 million infants annually for certain genetic, endocrine, and metabolic disorders, as well as congenital hearing loss and critical congenital heart disease. The public health goal is to identify and treat conditions that could severely affect the child's future health and even survival. In its announcement of the "Ten Great Public Health Achievements—United States, 2001–2010," the Centers for Disease Control and Prevention (CDC) included newborn screening, pointing out that "improvements in technology and endorsement of a uniform screening panel of diseases have led to earlier lifesaving treatment and intervention for at least 3,400 additional newborns each year with selected genetic and endocrine disorders."

Obviously newborn screening is important for those affected infants, but the odds of any particular infant being benefited is less than one in a thousand—meaning that Christina and Robert should not have to worry much about the results of the screening. As Stefan Timmermans and Mara Buchbinder conclude in their 2013 study of newborn screening in California, "Newborn screening may make a world of difference to individual families, but no available data has shown that newborn screening is associated with a reduction of infant mortality at the population level." Newborn screening, the authors note, is a public health anomaly, since it is not aimed at common diseases: "The irony of expanded newborn screening is that the United States instituted a public health program aimed at prevention for very rare conditions." How did this happen?

The model disease on which newborn screening was founded is phenylketonuria (PKU). It is a condition that is devastating if not diagnosed early, and the symptoms can be controlled by a restricted diet. In 1934, a Norwegian physician-chemist named Asbjørn Følling was consulted by the mother of two children, a brother and sister, with severe "feeblemindedness." He determined that the urine of these children had very high levels of a chemical called "phenylpyruvic acid." The condition eventually became known as phenylketonuria. In 1960, the microbiologist Robert Guthrie (who had a son and niece with developmental disabilities, the latter diagnosed with PKU) developed an inexpensive, sensitive, and simple bacterial inhibition assay, using a blood spot from the infant on special filter paper, that could be administered a few days after birth on a large-scale population basis. This test became known as "the Guthrie test."

In 1961, President John F. Kennedy (whose sister Rosemary was mentally disabled) promised to double the money spent by the National Institutes of Health on "retardation" research, and appointed a Presidential Advisory Commission on Mental Retardation, charging it with appraising the adequacy of existing programs. The commission hired the Advertising Council, which, among other things, mounted a dramatic campaign advocating that the new PKU test "should be a must for all babies everywhere." The National Association for Retarded Children was particularly influential in garnering political support and proposed model state legislation (because newborn screening is governed by state law) for mandatory PKU newborn screening. By 1975, forty-three states had enacted such laws and 90 percent of all newborns were being tested. Today, every newborn screening program in the United States includes PKU.

The mainstay of treatment for PKU is a strict diet with very limited intake of phenylalanine, which is mostly found in protein-rich foods. The goal is to consume only the amount of phenylalanine required for normal growth and body processes but no more. Because breast milk and regular infant formula contain phenylalanine, babies with PKU must have their diets substituted with a phenylalanine-free infant formula. It was once believed that it was safe for a person with PKU to stop the diet in adolescence, but it is now recommended that patients remain on a phenylalanine-restricted diet for life. On the other hand, long-term data is very spotty, with almost 70 percent of PKU patients lost to follow-up.

In addition to deciding which specific diseases warrant inclusion in screening panels, the major ethical issue with newborn screening is the informed consent of the parents. The vast majority of states make newborn screening mandatory, although (like childhood vaccinations) most have ways to opt out for religious or other reasons. Of course, you cannot opt out if you don't know about it in the first place. In 1982 Ruth Faden and her colleagues published a study of informed consent from Maryland, a state that has required informed consent for PKU screening since 1976. Their primary question was whether the parental consent requirement interfered with the public health mission of the screening. The conclusion was that it did not. In fact, of the 50,000 women studied, only 27 (or about 1 in 2,000) refused to consent, a number, the authors note, that is 100 times less than the chance of missing a PKU infant because of a false-negative result. In a companion article, the authors argue that consent should not be required because the central issue is child welfare, not parental autonomy. In the context of PKU screening, they ask, "Is a public policy that grants parents the right to consign their children to a state of irreversible mental retardation morally acceptable? We think not."

George took the other side, arguing that Faden's own data undermined her ethical argument. First, there really is no conflict here between the autonomy of parents making their own decisions (liberty) and the requirement that parents act in the best interests of their children (beneficence). With such a low rate of refusal, we can accommodate both values simultaneously. As George noted, at the observed rate of refusals (1 in 2,000), "it would take 500 years before one case of PKU is missed" because of parental refusals.

Second, parental refusal cannot in any reasonable way be construed as child neglect or denying your child an important medical benefit. This is because we know to a higher degree of certainty than we know about most other things in medicine that your child does *not* have PKU. This is because only one in 15,000 infants has this condition, making the odds "overwhelming that if any particular set of parents refuse screening, no detriment at all will befall their child." The real question is not about the parents but about the state: "Is mandatory screening for PKU a legitimate exercise of the state's public health powers?" The answer to that question "requires an analysis of testing and treatment technologies, the incidence of the disease, resource allocation, and the role of law in promoting the nation's genetic well-being."

George agreed that newborns should be tested for PKU based on these criteria (and so does Sherman) but argued that the Faden study demonstrated mandatory screening laws to be unnecessary to protect children. George also worried about the future of screening for hundreds of conditions, which would inevitably yield high numbers of false positives. George noted specifically that if we screen for one thousand diseases and each test is so good it has only a 1 percent rate of false positives, each infant will initially be diagnosed with ten diseases—even though the infant in fact has none of these—likely generating pathology from the retesting procedures and making refusal of testing in the first place a rational decision.

We have not yet gotten to that thousand-disease test threshold, but we are moving in that direction. Driven primarily by new technology that permits rapid assessment of many conditions via tandem mass spectrometry, in 2006 an ACMG task force specifically recommended that all state-based newborn screening programs adopt an expanded uniform panel of twenty-nine core (primary) conditions and twenty-five secondary conditions. Subsequently, two additional conditions were added to this list, severe combined immunodeficiency (SCID) and critical congenital heart disease (CCHD), for a total of thirty-one primary conditions.

All states now screen newborns for the recommended twenty-nine core conditions, and some now include one or both of the newest core conditions, SCID and CCHD. The number of secondary conditions included in newborn screening panels varies among states, as do requirements for parental education about newborn screening, consent and notification processes, payment for newborn screening, provision of care for affected children, laboratory standards, and storage, use, and disposal of blood specimens. For example, coverage for medical formulas needed to meet a child's metabolic requirements and foods needed for dietary treatment is a patchwork across states. States also finance their newborn screening programs in different ways. Most states collect a fee for screening, which differs depending on the disorders in the testing panel. For state-required conditions, private insurance or Medicaid usually covers the costs. Parents can also pay private laboratories for additional conditions, although there is almost never any reason to do this.

A number of professional organizations and government agencies have published guidelines for newborn screening based on three main considerations: (1) there should be a health benefit to the child from detecting the disorder in the newborn period, (2) the overall benefits and risks should be balanced by the financial costs, and (3) the harms should be minimal as measured by false assignment of positive or negative results. All of the guidelines emphasize that newborn screening is not just a blood test. It is a process that involves communicating timely information to parents, sampling, quality laboratory analysis, counseling when necessary, appropriate referral of an affected baby for the start of treatment, access to treatment, and follow-up evaluation of outcomes. The primary rationale for including a condition in a uniform screening panel is that it has a *direct benefit to the newborn child.*

Newborn screening has undeniable benefits, but only for a very small number of people. For example, early diagnosis of PKU and initiation and maintenance of dietary treatment can help prevent severe, irreversible brain damage and intellectual disability. In the United

States, about two hundred babies born with PKU are detected annually. But, as already suggested, there are real downsides associated with newborn screening. Perhaps the most vexing problem is a false positive, when an infant is identified as needing additional testing because the screening results are outside the normal range, but follow-up testing shows the infant is actually unaffected. A *minimum* estimate is that the false-positive rate for newborn screening is 1 in 300 infants. Assuming a U.S. birth rate of 4 million per year, this would translate into more than 13,000 infants having false-positive results. One study found that, *on average,* there are more than 50 false-positive results for every true-positive result identified through newborn screening in the United States. For any particular test, the rate may be even higher. For example, a study of maple syrup urine disease found that there were sixty-eight false positives for every true positive.

Johanna L. Schmidt and her colleagues studied the impact of false-positive newborn screening results on families. They interviewed parents with children whose follow-up test results indicated that the initial newborn screening result was a false positive. The stress, adverse parent-child relationships, and increased perception of child vulnerability these parents experienced are evident from three representative mothers:

> Upon receiving the initial positive newborn screening result, a mother (Helen) said, "I'm starting to get emotional here, but it was very stressful for me. . . . I was thinking, 'Oh, my gosh, is he going to die?'"
>
> During five follow-up retesting procedures of her child, a mother (Yolanda) described her experience: "The state kept on saying, you know, 'Okay, his levels were low, but they weren't low enough, come back.' So much that when [my baby] saw the nurses, he would just cry."
>
> Another mother (Katelyn) described her feelings during the follow-up testing: "So, when she pricked my baby's feet . . . it looked so painful. . . . He started to cry, and it's like, 'Oh my goodness, I'm just torturing my child, just for this.'"

For most parents, the anxiety they experienced from the time they learned the initial screening results continued until they learned that it was a false positive. One mother (Margaret) described the period as "a nightmarish two and a half weeks." A second mother (Elizabeth) described how nothing could calm her worry: "Nothing made me feel better. . . . If God had come down, if a prophet had come down from heaven and told me that he did not have cystic fibrosis, I would not have believed it until I got the sweat test."

Related to the problem of false-positive results is overdiagnosis, which occurs when confirmatory testing of a child with a positive screening result shows a biochemical abnormality or a gene mutation identical to that of a child with the disease, but the child never develops symptoms. When a physician encounters an *asymptomatic* child with abnormal laboratory results consistent with a disease, there is the dilemma of whether or not to proceed with treatment, especially if the disease is one that is known to sometimes, but not always, be life threatening or cause permanent harm. An example is 3-methylcrotonyl-coenzyme A carboxylase deficiency, which has outcomes ranging from asymptomatic (that is, a nondisease) to death.

An in-between situation occurs when an infant undergoes confirmatory testing (due to positive newborn screening results), and the follow-up tests are inconclusive. For example, the initial newborn screening for cystic fibrosis (CF) involves examining the dried blood filter paper for increased levels of immunoreactive trypsinogen (IRT), an enzyme produced by the pancreas. If positive, the next step is to look for mutations in the *CFTR* gene that result in an abnormal protein, which causes the clinical symptoms of cystic fibrosis. However, not all *CFTR* mutations cause disease, and clinical testing that entails measurement of sweat chloride values is not always clearly in the diagnostic range. Older individuals in these categories have been characterized as having "atypical cystic fibrosis" or "mild variant cystic fibrosis," but they present for diagnostic evaluation because of signs or symptoms, whereas infants identified by CF newborn screening are symptom free. For infants identified through newborn screening programs in which cystic fibrosis cannot be diagnosed or clearly ruled out, the term *CFTR-related metabolic syndrome* has been proposed. Whether these children will ultimately develop cystic fibrosis–like symptoms is uncertain. This inevitably results in numerous follow-up tests and "watchful waiting," where treatment is not given unless symptoms appear or change. This is stressful for the child and parents, as they wait for the *possibility* that the disease may appear.

Parental education about newborn screening needs significant improvement. Only about half of U.S. newborn screening programs provide parents with educational materials such as pamphlets or brochures, and content is inconsistent. If information about newborn screening is provided, it is usually given in the hospital shortly after delivery, when the parents are often emotionally and physically exhausted and focused on many other things. The American College of Obstetricians and Gynecologists sensibly recommends that education about newborn screening be provided during prenatal visits.

Genomic Sequencing of Newborns

Newborn screening controversies were highlighted in a report by the President's Bioethics Council in 2008, which concluded that the concept of benefit was being greatly and uncritically widened to include not just the child but the family and society itself. Reflecting on a future when genetic screening might be available for newborns, the council argued that harms involved in detecting disease susceptibility and diseases for which there is no treatment will be accentuated and not balanced by benefits: "Today's debates over whether to add this or that rare disorder to a uniform screening panel will be swamped, in the context of genomic medicine, by a radically more expansive approach to genetic screening."

Should we begin to think about replacing current newborn screening (largely based on the use of tandem mass spectrometry to measure various individual chemical compounds in the blood) with whole-genome sequencing? Since the underlying basis for these diseases almost always involves gene mutations, the question must be faced: would it not be better to have the entire genome sequence and thereby detect mutations in all genes at birth—or, as envisioned in *Gattaca,* even before pregnancy by preimplantation genetic diagnosis—rather than performing DNA analysis on a single gene or a series of genes each time a diagnosis is sought? Whole-genome screening would permit detection of diseases even rarer than those now on current newborn screening panels. The individual's genome sequence readout could then be entered once into his or her electronic health record and queried as needed for diagnoses, treatments, and prevention measures for the rest of the child's life. We must debate whether this would be a good thing, but like tandem mass spectrometry in the newborn screening arena, genomic sequencing technology and its cost will likely be the primary drivers in our technology-driven health care system.

The major reason for our arguably gloomy view is economics. Routine genome sequencing of potentially every baby born in developed nations provides a massive market for companies competing in this arena, and many companies will soon begin to market their WGS tests to countries such as China and individual states in the United States. Nonetheless, states may be able to resist the sales pitches, since the cost of follow-up will be many times the cost of the screening itself, and controversy already exists over what samples and what data of the newborn the state should be able to retain and use. Moreover, having the entire genome sequence of a newborn (or adult, for that matter) will likely not be all that valuable in terms of disease prediction, treatment, or prevention for individuals, at least not for the foreseeable future. There will be exceptions, such as finding mutations for serious conditions, including inherited cancer genes (a topic we cover in the next chapter). The critical issue is whether finding a variation (mutation) in the genetic code is medically *actionable*; in other words, does it translate into a clinically relevant action that can cure, ameliorate, or prevent a disease or other adverse impacts on health?

The data that will become available from genome sequencing is essentially endless, akin to the amount of information one can access on the Internet, and like the Internet, only a small portion of the information will be of value to any individual. Who should decide what we look for and what we disclose, and how, especially from newborn screening? The easy answers—search for everything but tell the parents nothing, or tell the parents everything—are both equally wrong. The challenge will be to find a reasonable middle ground, and this is unlikely unless we carefully consider the ethical, legal, and practical issues now.

Bioethicists Karen Rothenberg and Lynn Bush write plays as a pedagogical approach to genetics. The fictional characters in one suggest a possible future conversation on newborn screening: "Dr. Labgen offers with enthusiasm, 'Newborn whole-genome sequencing will help save more kids.' Having a different opinion, Dr. Pedethic counters, 'You've got to be kidding me. From what? There is no way whole-genome sequencing of newborns won't create a massive degree of patients in waiting.'" The committee chair concludes, "What we really need is good empirical data . . . where, when, and for whom it will help, and under what conditions it would cause more harm than good."

Anticipating this debate, and perhaps seeing routine newborn genomic screening as an potential grand entry way to introduce genomics as a central element of clinical use, NIH announced in late 2013 that it had funded four research projects on using WGS on newborns. Robert C. Green and Alan Beggs of Boston Children's Hospital lead one of them. Green's part involves recruiting the parents of healthy newborns who agree to have their child's genome sequenced, then providing information to the parents to see how they use it. It will not be possible to assess the outcomes for at least eighteen to twenty years, arguably longer. So we think the newborn researchers are very premature in concluding that "we are entering an era where all of medicine is genomic medicine." We think it is much more accurate and evidence based to say that "some" medicine will be genomic medicine, and it will not necessarily include newborn screening. We support research, but research on children who cannot give their consent can be done only with the meaningful consent of their parents, and parents can only legitimately provide consent if the risk to their children is minimal and the children can be protected from predictable harm, including stigma. The challenges of doing genomic research on healthy newborns will be to protect the child's right to an open future, and avoiding an early genomic designation that puts the child in the "sick" category for life.

We also agree with the assessment of the editors of *Nature* that newborn WGS research highlights five questions:

1. Should it be done when we don't know "with any certainty what a given genetic variant will mean for a given individual"?

2. Is it likely that providing parents with uncertain information, such as cancer risks, will lead to "years worrying about that cancer risk in their perfectly healthy child"?

3. Can clinical sequencing provide useful and timely information to parents?

4. "Who owns the genetic data?" (a topic we take up in chapter 9).

5. "Should the data be shared with other researchers?"

We also think the *Nature* editors were right to place ethics before science in their conclusion that when dealing with newborns, it is imperative that scientists get the *ethics and science* right" (emphasis added). We do not believe that state-mandated genomic sequencing of newborns is either desirable or inevitable. We should not seriously consider adopting even private genomic sequencing of newborns before it can be demonstrated, based on valid research, that it is likely to do much more good than harm, and that parents are capable of providing informed consent for themselves and their children.

Newborn screening has been sold based on the model of PKU disease. Public health in the United States strongly supports medical screening that might help even a few people. At the same time, American medicine tends to marginalize methods that could actually greatly improve the overall health of the large group of extremely premature infants in the United States. We rank very poorly in the world in terms of infant mortality, and the concentration on genetic screening of newborns will have no impact on this sad state of affairs. Nonetheless, we love our technology, and genetic screening, whether based on evidence or not, is quickly moving from looking for biochemical markers for newborn diseases to broadening out to the entire population to search for genetic markers for cancer. Cancer sells, and it is to cancer that we turn in the next chapter.

WHEN THINKING ABOUT GENOMIC SCREENING OF CHILDREN AND NEWBORNS, CONSIDER THESE THOUGHTS

WGS of children is in its infancy and is only medically indicated for children with serious, undiagnosed conditions.

WGS of sick children and newborns should never be done without parental consent.

Newborn screening is a public health tool for population screening.

WGS of healthy children and newborns should not even be seriously considered until research demonstrates that it is likely to produce much more benefit than harm to children.

Labeling a child genetically abnormal can transform a healthy child into a sick child for life.

CHAPTER 8

Cancer Genomics

Everyone who is born holds dual citizenship, in the kingdom of the well and in the kingdom of the sick.

—Susan Sontag, *Illness and Its Metaphors* (1977)

Much of the action in contemporary genomics relates to cancer. There is a good reason for this. Cancer has been the most feared disease in the United States for more than a century. Efforts sponsored by the federal government to find cures for cancer date from the establishment of the National Cancer Institute (NCI) in 1937. Cancer research intensified after President Richard Nixon's declaration of a "war on cancer" in 1971, and again in 2009 when President Obama announced he would double the amount of money the NCI could spend on cancer research, and arguably yet again in 2015 when President Obama proposed a new genomics research project to, among other things, cure cancer. In this chapter we report on both how we are doing in the genomic war against cancer and why curing cancer is so hard. An accurate description of the current state of genomic science and cancer treatment was made in *Nature:* "Personalized, 'precision' medicine for cancer is in a difficult time of transition." It is a time of great promise and great challenge.

We begin with a homage to Susan Sontag, who after her first experience with cancer wrote a powerful meditation on the language we use to describe it, pointing out that the most common metaphors are military ones. As she observed, cancer cells don't just multiply; they are "invasive," they "colonize" the body, setting up outposts ("micrometastases"). The body's "defenses" become overwhelmed, "scans" of the territory (the body) are taken, and treatment aims to "kill" the cancer. There is, Sontag writes, "everything but the body count." She argued (mostly unsuccessfully) that we should abandon this metaphor because it is dehumanizing, leading us to treat the patient's body like a battlefield, to set no limits in our attempts to destroy the enemy, and to place no cost limits on our efforts, no matter how futile.

Cancer is the second most common cause of death in the United States, exceeded only by heart disease. The average lifetime risk of a man developing cancer is slightly less than one in two; for women the risk is slightly higher than one in three. However, for any given individual, the risks vary depending on such factors as environmental exposures (for instance, smoking) and genetic susceptibility. Nearly one in four Americans will die of cancer. In 2014 there were more than 1.5 million new cancer cases and more than 500,000 cancer deaths.

Environmental factors (as opposed to hereditary factors) account for an estimated 75–80 percent of cancer cases and deaths in the United States. These include tobacco use, poor nutrition, physical inactivity, obesity, certain infectious agents, certain medical treatments, excessive sun exposure, and exposure to carcinogens (cancer-causing agents) that exist as pollutants in our air, food, water, and soil. Of these, tobacco smoking accounts for 30 percent of all cancers, and the combination of poor nutrition, obesity, and inactivity account for an additional 35 percent of the cancer burden. Environmental and behavioral factors are potentially modifiable. It has also been suggested that the majority of variation in cancers involving different sites, including esophagus, intestines, and stomach, are due to "bad luck," specifically "random mutations arising during DNA replication in normal, noncancerous stem cells."

Cancer Is a Genetic Disease

Although only 5 to 10 percent of cancers have a heritable component, cancer is a genetic disease, the most common of all genetic diseases. Cancer arises through mutations (changes in the sequence of DNA) leading to uncontrolled rapid growth of cells. We can divide these mutations into two types: *somatic mutations* (found in the cancer itself but not in the rest of the body's cells) and *germline mutations* (inherited and present in every cell throughout the body). Most cancers are the result of somatic mutations, which are induced by environmental exposures. Nearly all cancers originate from a single cell through sequential accumulation of multiple mutations that ultimately overcome the control mechanisms regulating cell growth. The descendants of such a cell continue to divide, called *clonal expansion*, and undergo further genetic changes into invasive, metastatic cancer.

The purpose of cancer screening is to discover a malignancy at an early stage, before the patient has any signs or symptoms, so that treatment can be initiated, optimizing the chance of a cure. Cancer screening falls into five general categories: (1) history and physical examination (for example, unexplained weight loss or swollen lymph nodes); (2) laboratory tests (including blood, urine); (3) imaging studies (such as X-rays, ultrasound, or magnetic resonance imaging); (4) invasive procedures (for example, colonoscopy); and (5) genetic tests (mutations in specific genes).

All screening tests have disadvantages or carry risks. Invasive procedures, such as a colonoscopy, can rarely cause damage, such as perforation of the bowel. A test can show a false-negative result, in which the test indicates there is no cancer when there really is. This can lead to delay in treatment and adversely affect prognosis. A test can also show a false-positive result, in which the test is abnormal but cancer is not present. Beyond causing anxiety, this often leads to more invasive testing, such as biopsies, which result in possible discomfort, risk (such as bleeding and infection), and costs.

Some screening tests are offered to individuals who have a known increased risk for certain cancers (known as a *risk factor*). These include people who have had a prior cancer, those who have a strong family history of cancer, or those who have been identified as carrying gene mutations that predispose to cancer (usually discovered during the evaluation of a family history of cancer). The American Cancer Society provides guidelines for the early detection of cancer on their website.

A goal of genomic testing is to usher in an era of personalized cancer medicine. One idealized view of the future has been described by the American Society for Clinical Oncologists. In their hypothetical case, a woman (we'll call her Joan) has a "routine" blood test during her annual physical, and within a few minutes the results come back as showing cancerous cells in her bloodstream. Joan is told that these cells "are an indication of an early-stage cancer that is developing somewhere in your body." She is reassured that since the cancer has been detected at an early stage, "there is a good chance that it can be managed or cured."

Joan is referred to an oncologist, who recommends additional genomic tests to determine the molecular "fingerprint" of the cancerous cells and the gene and protein abnormalities that may be driving her cancer. Within a few hours, the laboratory results are returned, indicating that Joan's cancer is developing in her kidneys. The oncologist says it's not the tumor's location that's important but her genomic profile and the unique combination of molecular features of her cancer. Based on Joan's medical history and genomic predispositions in her electronic health record (EHR), as well as treatment information about other patients like her, the oncologist determines that Joan would probably have an adverse reaction to one of the standard therapies. The EHR also identifies a clinical trial in which she could enroll.

After obtaining a second opinion, Joan enrolls in the trial, which includes two new drugs, "which are attached to a microscopic 'nanoparticle shuttle' that will deliver them directly to individual cancer cells, sparing healthy cells and minimizing side effects." Joan also receives a saliva reader that attaches to her smart phone and mobile applications that record her symptoms during the trial and automatically transmit this information to her EHR. Her smart phone also notifies her when it's time to take her medication and asks her questions about how she is feeling. It also alerts her to expect to feel fatigued and provides suggestions for managing side effects. Joan feels reassured that her health care team knows a great deal about her cancer and that she is making informed decisions about managing her cancer, "while continuing to work and live an active life." How close to reality is this futuristic vision of cancer diagnosis and treatment? To begin to answer this question, we must start with our current understanding of the causes of cancer.

The Hallmarks of Cancer

Cancerous tumors are complex tissues made up of malignant cells as well as normal cells that interact with and support the tumor. Therefore, to understand the biology of tumors, it is not enough to just study the malignant cells; rather, the entire "microenvironment" of the tumor must be examined. As normal cells evolve progressively to cancer, they acquire a succession of distinctive and complementary capabilities that enable the tumor to grow and spread. Douglas Hanahan, director of the Swiss Institute for Experimental Cancer Research, and Robert A. Weinberg of the Whitehead Institute for Biomedical Research at MIT have described these capabilities as eight "hallmarks of cancer" in one of the most widely cited papers in the scientific literature. Of these eight hallmarks, six are established hallmarks and two are so-called emerging hallmarks. They can be summarized as follows:

1. Sustaining proliferative signaling

Perhaps the most fundamental trait of cancer cells is uncontrolled stimulation of cell proliferation—that is, cell growth and division. This happens because cancer cells grow even when they are not receiving signals (messages) to grow. In other words, cancer cells behave as if a growth stimulus (called a growth "factor") were present even when it is not. They are no longer dependent on growth factors; they become "masters of their own destinies."

2. Evading growth suppressors

Most normal cells have a series of "brakes" that prevent them from continuing to grow and divide. Cancer cells lose their ability to respond to these "antigrowth signals," which come from tumor suppressor genes.

3. Resisting cell death

Normal cells have a built-in mechanism for programmed "suicide," termed apoptosis. For example, if a skin cell accumulates excessive DNA damage from too much exposure to UV light, it has mechanisms to self-destruct. Cancer cells have ways to disrupt the cell death mechanisms that carry out the suicide mission. Instead, cancer cells keep on growing and dividing.

4. Enabling replicative immortality

At the tips of chromosomes are stretches of DNA sequences called telomeres. Telomeres are made of a few thousand repeating sequences of TTAGGG. In normal cells, the telomeres limit the number of times a cell can divide. Each time a cell divides, the telomeres get shorter, in the range of ten to thirty-five TTAGGG repeats. Telomeres have therefore been compared to a fuse on a bomb. When the telomeres get too short (usually after fifty to seventy cell divisions), the cell can no longer divide, and it becomes inactive ("senescent") or dies. By contrast, cancer cells develop ways to lengthen their telomeres, thereby allowing them to divide an infinite number of times. In this way cancer cells achieve a form of immortality.

5. Inducing angiogenesis

Like normal organs, cancerous tumors require blood vessels to provide oxygen and nutrients and dispose of carbon dioxide and wastes. Most tumors develop signaling mechanisms to switch on the growth of new blood vessels, called angiogenesis, to keep up with their expanding needs. These new blood vessels are usually abnormal: convoluted with excessive branching, enlarged, and having erratic blood flow and leakiness.

6. Activation of invasion and metastasis

As cancerous tumors develop into higher grades of malignancy, they invade local tissues and spread to distant locations, called metastasis. These cancer cells change in shape as well as their ability to attach to other cells and to the extracellular matrix (ECM), the meshwork that surrounds cells and organizes them into structures. These capabilities of cancer cells involve a multiple-step process of mutations in genes that encode molecules for cell-to-cell and cell-to-ECM invasion, the invasion-metastasis cascade.

7. Emerging hallmark: Reprogramming energy and metabolism

Cancers need to make adjustments to rapidly increase or redirect their energy metabolism (for example, usage of glucose) to fuel cell growth and division. This reprogramming of energy and metabolism is largely orchestrated by proteins that are involved in one way or another with the six aforementioned hallmarks.

8. Emerging hallmark: Evading immune destruction

There is an increasing body of evidence from animal and clinical studies suggesting that the immune system acts as a significant barrier to tumor formation and progression. It is thought that cells in at least certain types of cancer may evade destruction by disabling components of the immune system that have been dispatched to kill them.

Scientists are currently developing targeted therapies aimed at one or more of these hallmarks (figure 8.1). The sobering reality is that tumors seem to be able find a way around these targeted therapies, and cancers therefore often don't stay in remission forever. The hope is that if multiple hallmarks are targeted simultaneously, a tumor won't be able to get around all these barriers at the same time.

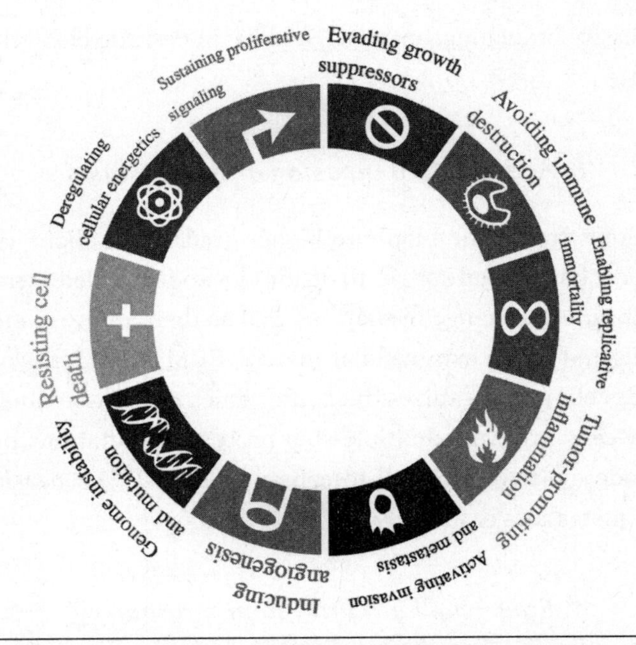

8.1 The Hallmarks of Cancer, D. Hanahan and R.A. Weinberg, "Hallmarks of cancer: the next generation," *Cell* 144 (2011): 646-74.

Angelina Jolie Pitt is probably the best-known cancer patient in the United States, though, as we discussed in chapter 1, she has never had cancer. Americans like to read about celebrities, but they don't like to read about dying celebrities. Nonetheless, the well-known acerbic and hard-living writer, Christopher Hitchens, famous for writing a nasty book trashing Mother Teresa, *The Missionary Position*, was determined to write about his own experience in cancer treatment. His story has special meaning for us because his progress was closely followed by NIH director Francis Collins (when you are a celebrity, you can get the attention of the head of the National Institutes of Health), who tried to come up with as many genomic treatment possibilities for Hitchens as he could. Hitchens wrote about his experiences in *Vanity Fair*, and the articles were later collected in his book *Mortality*. Hitchens, a long-time smoker and heavy drinker, had esophageal cancer. He describes his experience: "I have more than once in my time woken up feeling like death," but this time was different; he could barely breathe, and it took all his strength to call for emergency personnel. Thinking back, he found them courteous and professional as they escorted him, "a very gentle and firm deportation, taking me from the country of the well across the stark frontier that marks off the land of malady." In the emergency room, he was told his "next stop would have to be with an oncologist." He describes his new citizenship in "the sick country" as having some advantages: "The new land is quite welcoming in its way. Everybody smiles encouragingly and there appears to be no racism. . . . [A]s against that, the humor is a touch feeble and repetitive, there seems to be almost no talk of sex, and the cuisine is the worst of any destination I have ever visited."

He is informed that the cancer had spread to his lymph nodes and that they are "palpable" from the outside; he would get biopsy results in a week. Eventually Hitchens gets settled into a regime of chemotherapy and becomes philosophical, noting that the "absorbing fact about being mortally sick" is that you spend a lot of time preparing to die "while being simultaneously interested in the business of survival." In one of our favorite phrases, he writes, "This is a distinctly bizarre way of 'living'—lawyers in the morning and doctors in the afternoon—and means that one has to exist even more than usual in a double frame of mind."

A well-known atheist, Hitchens writes about all the believers supporting him. The "best of the faithful," he writes, is Francis Collins, whom he knows both for his genome work and "from various public and private debates over religion." Collins "has been kind enough to visit me and to discuss all sorts of novel treatments, only recently even imaginable, that might apply to my case," although later he writes, "In Tumortown you sometimes feel that you may expire from sheer *advice*." Hitchens is encouraged to explore a new "immunotherapy protocol" being done at NCI, which involves removing T cells from the blood, subjecting them to genetic engineering, and then reinjecting them to attack the cancer cells. There is a catch: his tumor cells have to "match," they have to express a protein called NY-ESO-1, and his immune cells have to have the molecule HLA-A2. His tumor has the protein but not the HLA match. Other trials are said to be under way, but he writes, "I am in a bit of a hurry, and I can't forget the feeling of flatness that I experienced when I received the news." Later, when *60 Minutes* runs a story of tissue engineering used to create a replacement esophagus for a man with his cancer, he immediately contacts Francis Collins, "who gently but firmly told me that my cancer has spread too far beyond my esophagus to be treatable by such means."

Realizing it's a very long shot, Hitchens decides to have "my entire DNA 'sequenced,' along with the genome of my tumor." Francis Collins picks up the story at this point, writing to Hitchens that if both sequences are done, "*it could be clearly determined what mutations were present in the cancer that is causing it to grow*. The potential for discovering mutations in the cancer cells that could lead to a new therapeutic idea is uncertain—*this is at the very frontier of cancer research right now*" (emphasis added). Yes, all these new technologies might help, but they came too late for Hitchens, who deteriorated quickly and died on December 15, 2011. Two of his last written lines: "Body turns from reliable friend to more neutral to treacherous foe . . . Proust?" and "Banality of cancer. Entire pest-house of side-effects. Special of the day."

Cancer-Genome Sequencing

The emergence of massively parallel sequencing (MPS), sometimes referred to as "next-generation sequencing," has been a major technological advance used in cancer-genome sequencing, the one Christopher Hitchens was hoping to take advantage of in 2011. MPS refers to a group of so-called high-throughput DNA-sequencing technologies, meaning that they can rapidly sequence DNA on the hundreds of gigabase (1 billion bases) scale. This has enabled whole cancer genomes to be sequenced on large sample sizes of tumor types. It has also made it possible to use a single technology to perform many kinds of genome analysis—for example, discovering single-point mutations (that is, a change of one base pair in the DNA sequence), assessing copy number alterations (an abnormal number of copies of one or more sections of the DNA), and detecting DNA methylation (a method by which gene expression is regulated) and foreign DNA (such as from viruses).

Bert Vogelstein and colleagues from Johns Hopkins University have noted that cheaper and more accurate genome-sequencing technologies are rapidly changing our fundamental understanding of cancer. The major points are that many genes are mutated in a typical cancer (in common solid tumors, such as those originating from the colon, breast, brain, or pancreas, an average of thirty-three to sixty-six genes show mutations not found in the individual's noncancerous cells), and like snowflakes, no two cancer genomes are alike. Tumors evolve from benign to cancerous by acquiring a series of mutations over time; some mutations, called "driver mutations," affect cellular signaling and regulatory pathways, each containing multiple genes.

There are over two hundred forms of cancer, and each type has its own genetic landscape. In 2006 the National Cancer Institute and the National Human Genome Research Institute launched the Cancer Genome Atlas (TCGA) to systematically map these genetic changes. Investigators from around the world sent normal and malignant tissue samples from patients to the TCGA researchers, up to five hundred samples for each tumor type. The cancer types were selected for their poor prognosis and overall public health impact. DNA, protein, and RNA were extracted from the samples, and each type of cancer was extensively characterized. The data has been made freely available to cancer researchers worldwide. The TCGA was declared completed at the end of 2014, but the NCI has stated that it will continue intensive sequencing in three cancers: ovarian, colorectal, and lung. Plans for the new studies call for including clinical data about each patient as well.

A frequent characteristic of cancer cells is that their nuclei are enlarged, distorted, and darkly stained with clumping; sometimes their appearance is described as "bizarre." The reason for this appearance is that the nucleus of a cancer cell often contains major chromosomal abnormalities, namely too many chromosomes. Another type of chromosomal abnormality frequently seen in cancer cells is a "translocation," where a piece of chromosome breaks off and sticks to another chromosome. The most common type of translocation in cancer cells is "reciprocal translocation," when pieces of two chromosomes exchange places. Most solid tumors have dozens of translocations. However, the majority of translocations appear to be passengers rather than drivers, because their breakpoints occur in "gene deserts," parts of chromosomes that have no known genes.

The textbook model of tumor development is a "Darwinian" competition whereby driver mutations enhance a cell's "evolutionary fitness," promoting outgrowth of that clone and progression toward cancer. Genome-wide sequencing studies of several thousand tumors have shown only about two hundred common driver genes. Of these, about half are tumor suppressor genes—genes that normally prevent cell growth and proliferation but mutate and stop working. The other half are oncogenes, genes that normally promote cell growth and proliferation, and have the potential to become hyperactive. Although additional mutation driver genes will undoubtedly be discovered, they will likely be in uncommon tumor types that have not yet been studied in detail. The acquisition of driver mutations is gradual and occurs cumulatively over years to decades.

Cancer Treatment Is Evolving

As we discussed in chapter 3, personalized or precision medicine has been one of the most hyped "revolutions" in modern medicine, and nowhere has the hope for its success been greater than for cancer. For over half a century, the mainstay of cancer treatments beyond surgery has been the use of chemotherapy drugs and radiation treatments that attack all fast-growing cells, killing those that normally grow fast (like white blood cells or cells that line the intestines) as well as the cancerous ones. In essence, this is a one-size-fits-all approach to cancer treatment. By contrast, personalized cancer medicine uses specific genomic information about a person's tumor to help establish a diagnosis, plan treatment, find out how well treatment is working, or make a prognosis. In broader terms, personalized genomic medicine takes into account both the tumor and the host—that is, genetic variations within the tumor (somatic mutations) and variations in normal tissues (germline mutations).

Genomic information about tumors helps in selecting an effective therapy and, equally as important, avoiding ineffective treatments. Inherited differences in how drugs are absorbed, metabolized, distributed in various tissues, and excreted can play a critical role in establishing proper chemotherapy dosing and staying clear of treatments with unacceptable risks of adverse drug effects (see chapter 5). Focused profiling of tumor DNA paired with anticancer drugs is increasingly being considered part of routine cancer care. For example, in 2014 the FDA approved a new drug for recurrent ovarian cancer called olaparib, together with a companion diagnostic test for specific mutations in the *BRCA* genes that would determine which patients were most likely to benefit from the new drug. There are now freely available online personalized cancer medicine resources, such as *My Cancer Genome* and the National Comprehensive Cancer Network, which serve as decision-making tools for physicians, patients, and researchers, providing up-to-date information on what mutations make cancers grow and related therapeutic implications, including available clinical trials.

Steve Jobs, cofounder of Apple Computers, provides an example of the role and limitations of cancer genomics. President Obama called Jobs "among the greatest American innovators—brave enough to think differently, bold enough to believe he could change the world, and talented enough to do it." Jobs was one of the wealthiest people in America, with an estimated net worth of $7 billion. In 2003, Jobs had a CT scan and other tests, which determined he had a rare form of pancreatic cancer, an islet-cell neuroendocrine tumor. Such tumors arise from the specialized cells within the pancreas that produce insulin, which controls blood glucose. His physicians urged him to undergo surgery to remove the tumor, but instead Jobs chose a variety of alternative remedies, including a vegan diet, juices, herbs, and acupuncture. Nine months later, the tumor continued to grow. Only then did Jobs agree to surgery and chemotherapy.

According to an authorized biography, *Steve Jobs* by Walter Isaacson, Jobs personally led his doctors on an aggressive scientific approach to the treatment of his cancer. He was one of twenty people in the world at the time to have all the genes of his cancer and normal DNA sequenced. The sequencing cost $100,000 and was done through a collaborative effort among Stanford, Harvard, Johns Hopkins, and the Broad Institute of MIT. Based on the unique molecular signature of Job's tumor, his doctors tailored his chemotherapy treatments. Jobs told Isaacson that he was either going to be the first "to outrun a cancer like this" using genomic sequencing technology or among the last "to die of it." Jobs died on October 5, 2011, at age fifty-six, two months before Christopher Hitchens died. The cancers that caused the deaths of Hitchens and Jobs were relatively rare. We now turn to two more frequent cancers that have been especially prominent in genomics research: skin cancer (melanoma) and breast cancer.

Melanoma

Overall, the lifetime risk of developing melanoma skin cancer is about 2 percent (one in fifty) for whites and about 0.1 percent (one in a thousand) for blacks. This translates to about 77,000 new cases of melanoma diagnosed and about 9,500 people dying of melanoma each year. It is one of the most common cancers in young adults, especially women. The rate of melanoma has been rising for at least thirty years and now accounts for about 5 percent of all new cancers. Patients with early-stage melanoma can usually be treated successfully with surgical removal, although the disease can spread widely. The primary treatment for patients with metastatic melanoma has been standard chemotherapy. The side effects of these drugs are highly toxic, including severe nausea, anemia, and hair loss. In the past, the prognosis for patients with metastatic melanoma was very poor, with median survival rates well under one year.

Historically, melanoma has been classified based on clinical characteristics like the thickness of the primary tumor, the number of cells dividing in a biopsy specimen, and whether there are ulcerated lesions. In recent years, however, it has become clear that the genomic makeup of melanoma tumors has important therapeutic implications. It is now known that about half of melanomas carry mutations in a gene called

BRAF. (BRAF is the awkward acronym for serine/threonine-protein kinase B-Raf.) Up to 90 percent of the mutations in the *BRAF* gene affect just a single DNA base change of a T to an A (called the BRAF V600E mutation). In turn, this change in the gene's DNA sequence results in substitution of the amino acid valine for another amino acid, glutamic acid, in the BRAF protein. The BRAF protein helps transmit chemical signals from outside the cell to the cell's nucleus. It plays an important role in one of the signaling pathways, known as the MAPK (short for mitogen-activated protein kinase) pathway. The MAPK pathway helps control cell proliferation, differentiation, movement (migration), and apoptosis (cell death). When the BRAF V600E mutation occurs, it disrupts normal functioning of the MAPK pathway; melanoma cells containing mutant BRAF disrupt MAPK signaling, leading to unbridled growth and survival.

A study published in 2010 involved thirty-two patients whose metastatic melanomas were unresponsive to standard therapy. DNA analysis of their tumors showed that all had the BRAF V600E mutation. These patients were given an orally administered drug, PLX4032 (later named vemurafenib), which specifically inhibited the BRAF protein made by the defective gene. Remarkably, this BRAF inhibitor induced complete or partial tumor regression in 81 percent of the patients and symptom improvement in the majority of patients. In some cases, lesions that had spread to liver, bone, and lungs seemed to disappear within weeks of the first dose. An accompanying editorial, entitled "Melanoma—an Unlikely Poster Child for Personalized Cancer Therapy," exclaimed that "these results represent a major breakthrough and provide proof of principle that the treatment of metastatic melanoma can be individualized for a substantial percentage of patients."

Confirming reports soon followed. Patients with the BRAF V600E mutation treated with vemurafenib or other BRAF inhibitors showed high and rapid response rates. Although side effects were common, they were generally not severe. Unexpectedly, however, about 25 percent of patients treated with BRAF inhibitors developed another type of skin cancer, squamous cell carcinomas. These lesions could usually be managed successfully by local surgical removal and did not require discontinuation of BRAF inhibitor therapy.

Then bad news started to come in. Responses were mostly partial, and about half of patients began having cancer progression within six to seven months of treatment, as the tumors became resistant to the BRAF inhibitors. Inhibiting BRAF activity alone would not eradicate or even hold back melanoma tumors for very long. A number of new strategies are being tried. For example, combining a BRAF inhibitor with an inhibitor of a second enzyme in the MAPK pathway, called MEK (mitogen-activated protein kinase), helps fight drug resistance. Combining inhibitor drugs within and across signaling pathways, which target the molecular makeup of an individual's cancer, holds real promise in the treatment of melanoma as well as other cancers.

Another type of personalized cancer treatment, called "immunotherapy," attempts to target specific cancer cells by identifying proteins on their surface encoded by their unique genes. Ways to unleash the power of the body's own ("personal") immune system on cancers have long been sought. The origins of modern immunotherapy for cancer can be traced back to the late nineteenth century, when William Coley, a young New York surgeon, began injecting live or inactivated bacteria into patients with cancer. Inoculating cancer patients with bacteria made sense, given the evolving understanding of the power of the body's immune system to cause inflammation and destroy invading bacteria by stimulating antibacterial white cells that might kill bystander tumor cells. "Coley's toxin" was used to successfully treat a wide variety of cancers, including melanoma, sarcomas, carcinomas, and lymphomas; complete, prolonged regression of advanced malignancy was documented in many cases. Use of Coley's toxin was nonetheless opposed by the medical establishment because of mixed results. The modern science of immunology has shown that Coley's principles were correct, and Coley is now known as the "father of immunotherapy."

There have been intensive efforts to develop immunotherapeutic approaches to cancer. This is no easy task, because cancers have complex ways of escaping immune detection. For instance, tumors can decrease the expression of proteins on their surface, rendering them invisible to certain types of white blood cells (called cytotoxic T cells) that destroy target cells. Some tumors secrete proteins that actively suppress antitumor immune responses. Nonetheless, advances have been made in cancer immunotherapy. It has been observed that the "best approach to treatment might be to combine precision [genomic] therapy with immunotherapy." The thought is that genomic-based interventions can provide a short-term fix, whereas immunology approaches, when they work, are effective for much longer periods of time.

One strategy that has been used in melanoma and other cancers involves a gene called *PDCD1,* which encodes for a protein called PD-1 (short for programmed cell death protein 1). PD-1 is a receptor on the surface of T cells that interacts with a molecule called PD-L1 (short for programmed cell death 1 ligand). Normally PD-L1 fits with PD-1 like a lock and key to maintain the balance of the immune system by shutting it down at appropriate times. This is called an "immune checkpoint." Some melanomas (and other cancers) take advantage of this shutdown mechanism by continuously producing PD-L1, enabling the cancerous cells to escape being destroyed by T cells.

The drug pembrolizumab (formerly lambrolizumab) is a highly selective antibody against PD-1 that blocks the cancer cells from turning off the immune regulatory signaling of the PD-1 receptor expressed by T cells. This blocking antibody enables T cells to recognize proteins on cancerous cells as "foreign." The antibody doesn't fight the cancer directly, but it allows the body's own immune system to target cancer cells for destruction. In a clinical trial of 135 patients with advanced metastatic melanoma who were treated with lambrolizumab, 38 percent improved. Among those who received the highest dose, 52 percent improved and 10 percent had a complete response, meaning their tumors could no longer be detected on scans. Side effects were mild and easily managed. A second clinical trial showed comparable results "with rapid and deep tumor regression in a substantial proportion of patients." The results of these two clinical trials were heralded as "striking" and described as taking immunotherapy for melanoma to the next level.

Ralph Steinman, former director of the Laboratory of Cellular Physiology and Immunology at Rockefeller University, is an example of the potential of immunotherapy. In 1973, Steinman and a colleague, Zanvil Cohn, discovered a new class of cells, known as dendritic cells, that directs and regulates the body's immune system by programming other cells to recognize and destroy intruders. In 2007, Steinman was diagnosed with advanced pancreatic cancer. With the help of dozens of collaborators around the world, he began running a series of research studies on himself, a kind of personalized immunotherapy. In all, he tried eight experimental treatments, each approved under a single-patient, compassionate-use protocol. Steinman used three vaccines, all based on his dendritic-cell research. The idea was to boost his own immune response to the cancer by inserting proteins from the surface of his tumor into the dendritic cells and injecting them back into his body. Survival for patients with Steinman's type and stage of pancreatic cancer is usually measured in weeks or months; he lived for more than four years. Which, if any, of his treatments extended his life is unknown, but the work that he and his collaborators did advanced the field by demonstrating that conventional chemotherapy could be combined with dendritic-cell vaccines.

On October 3, 2011, it was announced that Steinman was to share in the Nobel Prize in Physiology or Medicine. Hours after the announcement, the committee learned that Steinman had died three days earlier. Deceased individuals have not been eligible for the prize since 1974. The committee consulted with their lawyers, who interpreted the rules as not applying if the committee was unaware that the winner was dead. Like Steinman's work on immunotherapy, change in American medicine is not revolutionary but evolutionary. Nowhere is this truer than in the area of cancer. We all hope for major cures, so-called magic bullets that will take away or, better yet, prevent cancers in ourselves and our loved ones. Such "overnight" breakthroughs rarely happen, but personalized medicine is slowly taking hold.

Breast Cancer

In the United States, about 230,000 persons will be diagnosed with invasive breast cancer, and about 40,000 deaths will occur from the disease in 2015, assuming current trends continue. One in eight women born in the United States will develop breast cancer sometime before they die. Another way of saying this is that one in eight women in the United States *who reach the age of 80 can expect to develop breast cancer*—although this figure is misleading and includes cancers that will never be detected. In each decade of life, the risk of getting breast cancer is actually much lower than one in eight, although risk increases with age. For example, the ten-year risk for breast cancer is one in sixty-nine (or less than 2 percent) for a woman at age forty; it is one in forty-two (a little more than 2 percent) at age fifty, and one in twenty-nine (about 3 percent) at age sixty. These probabilities are averages for the whole population. An individual woman's breast cancer risk may be higher or lower depending on a number of factors. The benefits and harms of a fifty-year-old woman being screened with an annual mammogram each year for ten years (until she's sixty) are

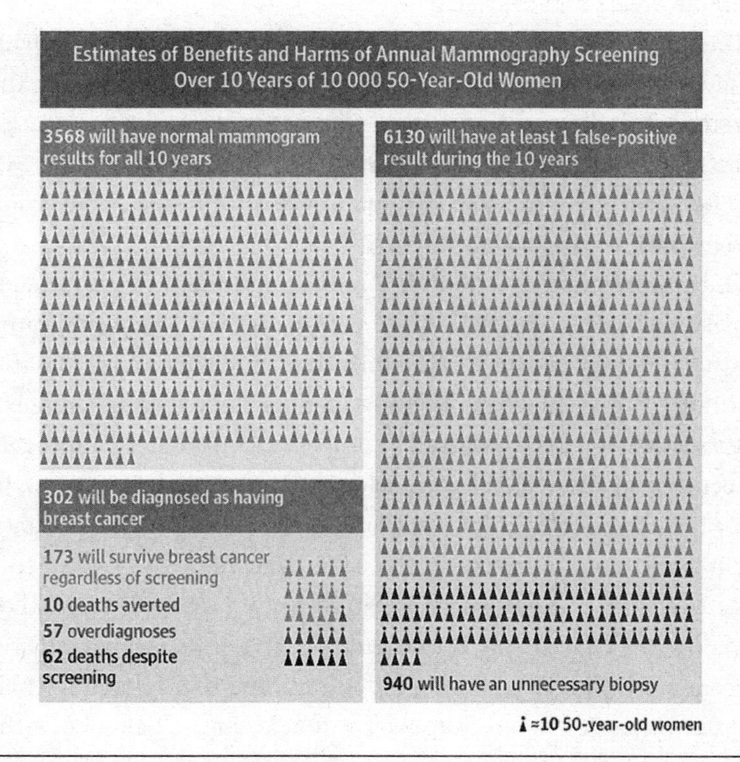

8.2 Visual aid to understanding breast cancer screening. J. Jin, "Breast Cancer Screening: Benefits and Harms," *JAMA* 312 (2014): 2585.

illustrated in figure 8.2. As seen, the woman will have more than a 60 percent chance of at least one false positive, 940 (or 9.4 percent) will get an unnecessary biopsy, fifty-seven will be overdiagnosed, sixty-two will die regardless of screening, and ten deaths will be averted. Only one woman in three will have a normal mammogram each year for the ten-year period. The bottom line is not that mammography should be discontinued for this age group, only that it helps postpone very few deaths (10), and six times as many breast cancer deaths (62) occur despite yearly mammography. Of course in the nondeath category, 940 will get unnecessary biopsies, 57 will get overdiagnosed and undergo unnecessary treatment, and 173 will survive breast cancer regardless of screening. Few clinicians, and fewer women, will disagree that we need a more effective approach to breast cancer prevention. Family history can provide a more effective approach for women at risk for heritable breast and ovarian cancer.

Diagnostic evaluation is usually undertaken because the patient or the health care professional feels a suspicious breast lump, or there is a suspect finding on a screening mammogram. Additional imaging studies using ultrasonography or magnetic resonance imaging (MRI) may be used. The next step is to obtain cells or tissue for diagnostic studies from the suspicious lesion.

Once breast cancer is found, a number of tests are performed to classify the tumor, determine the prognosis, and select the optimal treatment. Breast cancer is commonly treated by various combinations of surgery, radiation, chemotherapy, and hormone therapy. Important factors that need to be assessed to provide a personalized approach to cancer treatment include how quickly the cancer is likely to grow, how likely it is to spread throughout the body, how well certain treatments might work, and how likely the cancer is to recur.

After gender and age, a positive family history is the strongest known predictive risk factor for breast cancer. About 10 percent of breast cancers are linked to *germline* mutations, due to a hereditary mutation that is passed down from parent to offspring. Chief among them is the hereditary breast and ovarian cancer syndrome. Characteristics of this syndrome are early onset of breast or ovarian cancer (usually defined as younger than age forty) in multiple generations, individuals with cancers developing in both breasts or ovaries, Ashkenazi Jewish descent, and sometimes the occurrence of other cancers in family members (for example, prostate, pancreas, uterus, colon). Some of these families (some writers call them "cancer families," but it is wrong to reduce entire families to one characteristic they share) can be explained by single *cancer susceptibility genes,* with germline mutations in the *BRCA1* and *BRCA2* genes accounting for the vast majority of families with hereditary breast and ovarian cancer syndrome. The names *BRCA1* and *BRCA2* stand for breast cancer susceptibility gene 1 and breast cancer susceptibility gene 2, respectively.

The *BRCA1* gene is located on chromosome 17 and the *BRCA2* gene on chromosome 13. Both *BRCA1* and *BRCA2* are tumor suppressor genes encoding for proteins that function in the DNA repair process (discussed above). More than 2,500 distinct mutations have been described in *BRCA1* and *BRCA2*. In the general population, about 1 in 300–800 individuals carry a *BRCA1* or *BRCA2* mutation. About 3–5 percent of breast cancers and 10 percent of ovarian cancers are due to germline mutations in *BRCA1* and *BRCA2*. In the United States, about 1 in 40 individuals of Ashkenazi Jewish descent carry two specific *BRCA1* mutations (designated 185delAG and 538insC) and a *BRCA2* mutation (designated 617delT).

The lifetime estimated risk of a woman developing breast cancer ranges from 65 to 78 percent for *BRCA1* or *BRCA2* mutation carriers; the lifetime risk of a woman developing ovarian cancer ranges from 39 to 46 percent for *BRCA1* and 12 to 74 percent for *BRCA2* mutation carriers. Men with *BRCA2,* and to a lesser extent *BRCA1,* are at increased risk of breast cancer with lifetime risks in the range of 5–10 percent and 1–2 percent, respectively. Men carrying *BRCA2* mutations, and to a lesser extent *BRCA1* mutations, have about a three- to sevenfold increased risk of prostate cancer.

Of course, not every woman who carries a *BRCA1* or *BRCA2* mutation will develop breast or ovarian cancer. Even at a 78 percent risk, 22 percent, or almost one in four, will never develop the disease. Moreover, not every woman in families with hereditary breast and ovarian cancer syndrome necessarily carries a *BRCA1* or *BRCA2* mutation, and not every cancer in such families is necessarily linked to a harmful mutation of one of these genes. Evaluating a patient's risk for hereditary breast and ovarian cancer syndrome should be part of routine clinical practice. A number of clinical practice guidelines provide specific criteria clinicians may use in determining who should be offered genetic testing for *BRCA1* and *BRCA2* mutations and whether referral for genetic counseling is appropriate. Genetic counseling is generally recommended before and after genetic testing. The National Cancer Institute's recommendations for *BRCA1* and *BRCA2* testing are on its website.

A positive genetic test result for a *BRCA1* or *BRCA2* mutation has important health and social implications for carriers, other family members, and future generations. When evaluating a family, it is best, if possible, to begin with a person who has developed breast or ovarian cancer. Once a specific mutation has been identified in the affected individual, others in the family who choose to be tested can be studied for the same mutation. It is important to remember that both men and women who inherit a *BRCA1* or *BRCA2* mutation, whether or not they develop cancer themselves, can pass the mutation to their sons or daughters. There is a one in two chance of a mutation carrier passing the abnormal gene to a child. Conversely, there is a one in two chance that a mutation carrier will not pass the abnormal gene to a child.

If a person tests negative for a known mutation in his or her family, it is highly unlikely that they have inherited a susceptibility to the breast and ovarian cancer syndrome associated with *BRCA1* or *BRCA2*. However, this does not mean that they will not develop cancer; it means that the person's risk is probably the same as for anyone in the general population: one in eight. However, where there is a family history of breast and ovarian cancer but no mutation in *BRCA1* or *BRCA2* is identified, a negative test is considered "not informative." In other words, a harmful *BRCA1* or *BRCA2* genetic abnormality could exist, but it was not detected by the test. In addition, it is possible for people to have a mutation in a gene other than *BRCA1* or *BRCA2* that increases their risk of developing cancer but is not detectable by the tests used.

The recommendation to only test for cancer genes in healthy people if a close family member has been found to carry a particular gene was called into question in late 2014 with the publication of a study in

Israel that found women of Ashkenazi backgrounds have an unusually high incidence of *BRCA1* and *BRCA2* genes. While such screening is not currently recommended, many other "cancer genes" can now be screened for, with various panels of genes being suggested by different biotech companies. The quandaries these panels present to physicians and patients are well described by *Science* writer Jennifer Couzin-Frankel, who wrote about her own experiences (figure 8.3). Like many others, she read the *New York Times* article about the Israeli study in September 2014. Because both her parents were of Ashkenazi descent, she decided to consult a genetic counselor about *BRCA1* and *BRCA2* screening (the risks of carrying a *BRCA* mutation is one in eight hundred in the general population, and one in forty in the Ashkenazi Jewish population). After reviewing her family history, which included some possible cancers, the genetic counselor presented a list of twenty-one genes associated with breast and ovarian cancer, the genes on the "Breast/Ovarian Cancer Panel" marketed by a company called GeneDx

8.3 *Science* writer Jennifer Couzin-Frankel and her children. April Saul, in J. Couzin-Frankel, "Unknown Significance" *Science* 346 (2014): 1167.

in Maryland. As she described the table, eleven genes were shaded in pink and labeled "high-risk"; three in purple were "moderate-risk"; and seven in turquoise were described simply as "newer genes." Other companies recommended other genes. The *CHEK2* gene, colored purple, for example, could, she was told, double her risk of breast cancer. The impact of most of these genes on health remains uncertain—but as experts she consulted with (and had access to because of her profession) told her, the ability to test is moving much faster than our ability to interpret test results. As Kenneth Offit of Sloan Kettering Cancer Center told her, "This is the paradox we have fallen into." We are discovering more cancer genes at the precise time the cost of genomic sequencing is plummeting, with the result that the number of testing panels is proliferating."

In the article, which we recommend you read in its entirety, she describes her decision-making process, as well as her own tolerance for uncertainty, including whether she'd want to be informed of variants of uncertain or unknown significance. Ultimately she decides to have the panel of tests and to be informed of all the results, regardless of their known significance. She learns there are no variants of unknown significance in *BRCA*, but there is one in *CHEK2*. She is told that the latter variant had been found in two men with prostate cancer. Her response: "I expected distress, a ringing in my ears, fear coiling in the pit of my stomach. Instead, I'm almost laughing. I think, 'That's it? That's what is being shared with patients these days?' Two men with prostate cancer, cells in a petri dish, a loss of function that may or may not translate into pathogenicity: This does not merit my mental energy."

In the end she shares her results with her cousin, the only family member to whom it might matter. Her cousin urges her to get tested with a new panel—one for forty-eight genes, a test her cousin's mother just had because of her ovarian cancer (all negative). "In the end, I explain in my message to her, it wasn't something I wanted. 'I know the panels are often discouraged,' her cousin writes back. It's a view she doesn't share. Even without a clear-cut action plan, she wants to know whatever message her DNA carries for her future. The only reason she's eschewed testing for herself is because insurance is unlikely to pay for it."

Finding variations of uncertain significance is not unusual, even when testing is restricted to *BRCA1* and *BRCA2*. Overall, about 10–15 percent of individuals undergoing screening for *BRCA1* and *BRCA2* mutations will not find a clearly harmful mutation but will have a variant of unknown (or uncertain) significance (VUS). However, the proportion of individuals who receive a VUS result varies widely by population, with rates of up to 22 percent among Hispanics and 26 percent among African Americans. A VUS may cause substantial challenges in counseling in terms of cancer risk estimates, and clinical decision making must be highly individualized and take into account factors such as the patient's personal and family cancer history. Moreover, inherent in the implications of VUS is that they are moving targets. As additional data becomes available, a VUS may be reclassified as a benign variant or as deleterious.

Individuals who are identified as carriers of a harmful *BRCA1* or *BRCA2* mutation have a number of options to reduce their risks of developing cancer. These include:

- *Surveillance.* Surveillance refers to cancer screening with the goal of diagnosing the cancer early, when it is most treatable. Breast cancer screening may include mammography, clinical breast exams, and other breast cancer screening methods, such as MRI. While no specific type of surveillance has been shown to improve patient outcomes for those at high risk for ovarian cancer, some patients have opted for close monitoring using blood tests (for instance, CA125), transvaginal ultrasounds, and clinical evaluations. The effectiveness of such screening methods is uncertain.

- *Prophylactic surgery.* Bilateral mastectomy reduces the risk of breast cancer by 95 percent or more. Bilateral salpingo-oophorectomy (removal of both ovaries and fallopian tubes) has also been shown to reduce the risk of breast cancer by 40–70 percent. Bilateral salpingo-oophorectomy reduces the risk of ovarian cancer, fallopian tube cancer, and peritoneal cancer by 90 percent or more. Unfortunately, because not all "at-risk" tissue can be removed by these procedures, some women have developed breast cancer, fallopian tube cancer, ovarian cancer, or peritoneal cancer even after prophylactic surgery.

- *Chemoprevention*. Some breast cancers can be prevented by limiting cellular access to estrogen, which stimulates cell growth. Estrogen receptor blockers, such as Tamoxifen and Raloxifene, have been shown to reduce the risk of developing breast cancer in *BRCA1* and *BRCA2* mutation carriers.

The story of Ms. E (a real patient whom we'll call Emily) exemplifies how personalized cancer medicine is currently being practiced in the United States. Emily is a forty-one-year-old woman who works in the banking business. She is married and has had one child. At age thirty-seven, she began experiencing pain in her left armpit and noticed a lump that enlarged slightly over the next six weeks. She saw her gynecologist, who ordered a mammogram that showed only dense breast tissue, but MRI and ultrasound examinations revealed a mass in her breast as well as suspicious lymph nodes. A biopsy was taken of the mass, which showed an invasive ductal carcinoma of the breast. Emily was treated with wide surgical excision of the tumor and chemotherapy with three drugs (doxorubicin and cyclophosphamide followed by paclitaxel) and radiation therapy.

Because of her young age, Emily was referred to genetic counseling and underwent testing that showed she had a *BRCA1* mutation, more precisely a 538insC *BRCA1* mutation, which is one of the two common *BRCA1* mutations found among individuals of Ashkenazi Jewish descent. She was unaware of any Ashkenazi Jewish ancestry; however, her father was an adopted only child. In addition, the pathological studies performed on the biopsy tissue showed that her tumor was triple negative (ER-/PR-/HER2-), which is the case in 80 percent of *BRCA1*-related breast cancers. This was important information, because although recent data has suggested that an unconventional chemotherapeutic drug for breast cancer, cisplatin, might be useful for the treatment of newly diagnosed breast cancers in *BRCA1* mutation carriers, the standard therapy that Emily received was believed to be more appropriate than cisplatin therapy outside of a clinical trial.

Additionally, it is currently recommended that mutation carriers like Emily undergo prophylactic removal of the ovaries and fallopian tubes when childbearing is completed, ideally by age thirty-five to forty, to substantially reduce the risks of ovarian cancer and breast cancer. However, at the time of her diagnosis, Emily and her husband told her oncologist that they were trying to have a child through in vitro fertilization. Emily's eggs were harvested before she began chemotherapy. Subsequently, preimplantation genetic diagnosis was performed and an embryo free of the *BRCA1* mutation was implanted into Emily's uterus. (See chapter 4 for further details about preimplantation genetic diagnosis.) Four years after the initial diagnosis of her breast cancer, Emily gave birth to a healthy boy. Emily was quoted as saying, "The results of the genetic test have been extremely important for us. The idea of knowing is so much more important than not knowing. For us to have children at this stage in life and understand that we are not passing the gene outweighs any possible negative consequences."

A Final Perspective

What can we conclude from our exploration of cancer and the genome? All cancers arise from changes in the DNA sequence of normal cells. We have entered an era in which genomic profiling has begun to significantly improve the treatment outcomes for some patients with cancers. Importantly, some of the new drugs developed from what has been learned about the molecular biology of any one cancer, for example melanoma, hold promise in targeting pathways and checkpoints in other malignancies, such as colorectal carcinoma and kidney cancer. Still, we have a long way to go. As Frances Visco, the president of the National Breast Cancer Coalition, put it in the spring of 2015, "Having the technology to do genomic analysis of tumors does not mean that we have 'precision' or 'personalized' medicine," or that we can better help people with cancer. We agree with her that progress in cancer genomics must be measured not simply by developing "a new tool but rather stopping cancer deaths."

You and your family will increasingly encounter genomics in everyday health care, and screening for cancer or for a predisposition to develop cancer will likely play a prominent role. In the context of "personalized medicine," this will involve adjusting each person's level of risk rather than the average population risk. For example, men whose genomic profile show a lower than average risk for dying of prostate cancer may be able to avoid screening with prostate-specific antigen (PSA) with its inherent high false-positive rate, which leads to unnecessary biopsies.

We can now foresee a time, realistically as soon as a decade from now, when we will look back on the ways we now diagnose, treat, and try to prevent cancer as a kind of dark ages. As Siddhartha Mukherjee put it in *The Emperor of All Maladies*, "Arguably [the] most complex, new direction for cancer medicine is to integrate our understanding of aberrant genes and pathways to explain the *behavior* of cancer as a whole, thereby renewing the cycle of knowledge, discovery, and therapeutic intervention." With apologies to Susan Sontag, who might nonetheless agree with our conclusion if not our metaphor, the war on cancer has gone on far too long and has had far too many casualties. The peace treaty will be written in the language of our genomes and the genomes of the cancers that afflict us.

Further advances in cancer genomics will require sequencing hundreds of thousands, if not millions, of genomes so that relevant pathological genetic variants can be distinguished from irrelevant or harmless ones. This massive research project will require developing massive genomic data banks, and it is to the development of these data banks, and the privacy protections needed to encourage people to donate their genomes to such data banks, that we turn to in the next chapter.

WHEN THINKING ABOUT CANCER GENOMICS, CONSIDER THESE THOUGHTS

Cancer is a genomic disease that usually has an environmental trigger but can also be the result of just plain bad luck.

The "hallmarks" of cancer suggest specific components in cancer cells that can be targeted for treatment.

Sequencing cancer genomes holds promise for personalized medicine.

Immunotherapy is a fast-growing approach to cancer treatment that can be applied alone or in combination with genomic approaches.

Genomic screening for cancer can increase uncertainty by identifying variants of uncertain significance.

Breast cancer screening by mammography illustrates some of the problems of false positives and "overdiagnosis" that genomic screening will also create.

Learning more about cancer genomics will require massive genomic data banks.

CHAPTER 9

Genomic Privacy and DNA Data Banks

'What your DNA is telling you' doesn't just depend on your genome or the thousands of other genomes out there. Rather, it depends on how they can be related to one another; these vast amounts of data can be made sense of only by using computers to search for specific patterns and relationships.

—Hallam Stevens, *Life Out of Sequence* (2013)

The phrase *genomic privacy* suggests both that privacy is important and that there is a special kind of privacy we can label genomic. We think these assertions are true, and the goal of this chapter is to explain why and to suggest ways you can protect your own genomic privacy. Protecting your genomic privacy is made much more difficult than it should be in the current Internet age of Google and Facebook, as well as the National Security Agency (NSA), entities that often simply assert that "privacy is dead" and we should accept its death. Governments, private corporations, and researchers all want (and often get) access to our most personal and private information—sometimes with our active cooperation. Most Americans, however, disagree. Privacy is critical to human dignity and liberty. Most Americans believe that privacy should protect "private" information (especially medical and genetic information) to which we don't want others to have access without our authorization. In 2015 *Science* published a special issue on "The End of Privacy" that suggested not so much that massive genome banks make privacy obsolete, but that we will have to protect our privacy in new ways. President Obama underlined this concern when he assured the American public about his proposed one million person genomics bank that "We're going to make sure that protecting patient privacy is built into our efforts from Day 1." Privacy is critical not because we necessarily have "something to hide"; it's that we see no reason to share intimate details of our lives, including our medical information and our genomic information. Privacy also protects our identity. Privacy can be seen as making liberty possible, permitting us to make personal decisions that might otherwise be closed to us.

Genomic information is uniquely private because it implicates not just one or two but three major aspects of privacy: informational privacy, in that it is uniquely tied to you and your probabilistic future (what we have termed your "future diary" or your "personalized threat matrix"); relationship privacy, in that it discloses medical information about your parents, siblings, and children; and decision-making privacy (also called autonomy), in that it directly affects decisions about whether to attempt to become pregnant or to continue a pregnancy. Genomic information also has a history of misuse for eugenics in pre-World War II United States and genocide in World War II Germany. Finally, genomic information is not only stored and manipulated digitally (in a "DNA data bank"); it can also be stored physically, in the form of a tissue sample (in a DNA bank).

Since 9/11, the NSA has radically increased its power to spy on Americans, and Edward Snowden's public disclosure of unchecked data collection by the NSA has at least sparked public discussion of whether our government has gone too far in secretly collecting private information from ordinary citizens not suspected of any criminal activity. In genomics, the largest "gene banks" in the United States are those maintained by law enforcement agencies to identify criminals, or at least suspects. Have we gone too far in authorizing our government to take our DNA upon arrest? Should you want to have your DNA included in a collection of DNA samples in a massive data bank used only for medical research? How can we be sure that our DNA is not misused to our detriment by government and private entities? Should you care about DNA ownership, consent, collection, computerization, and information disclosure policies? Can you protect your genomic information and simultaneously let researchers use it to promote society's health? Should revelations about how large data collectors, including Facebook, have performed research on their users affect your views on the necessity of informed consent prior to "big data" research?

In this chapter we set out to answer these questions by first looking at the growth of DNA data banks, most notably government criminal DNA data banks, and the risks they pose to our privacy. These questions are not just hypothetical: they have been the subjects of major legal cases in the U.S. Supreme Court and the European Court of Human Rights. Comparing government DNA databases to private DNA databases, we use the second aspect of genetic privacy, family relationships, to see how our genetic privacy can be protected by private data bankers. The core question in this chapter is, who owns (in the sense of who can control access to and use of) your DNA? A series of real stories has played out in U.S. courts on this question, and we tell those stories. Should you view your DNA as your property or simply as highly sensitive medical information that should be protected by privacy laws? We will suggest privacy rules for DNA data banks, including the role that "depositors" in such banks should play in deciding who can have access to the data and whether it can be used for research or commercial purposes. We end the chapter with the remarkable U.S. Supreme Court decision that breast cancer genes are not patentable because they occur in nature, discussing the implications of that decision—the decision that corporations cannot own our DNA—for genomics, including whole-genome sequencing, and for our genetic privacy.

Big Data and Big DNA Data Banks

The core paradox of genomics is that while it promises "personalized medicine" based on treating you in unique ways informed by your unique genome, in the near future this will not be possible because not enough is known about population genomics. We need to conduct large-scale genomic research, research on the population level involving millions of genomes, to obtain the information we will need to make useful clinical decisions informed by genomics. Setting up these large-scale genomic research data banks is an ongoing project. The birth and growth of genomic data banks has concerned us for more than two decades, long before the era of big data, when the usual term was "DNA data banks." In 1992, this was the second on our list of the four most important legal and ethical issues raised by the new genetics: How can the privacy of an individual's genetic information be preserved? We argued that protecting privacy was critical for genomics because genetic information can be used to stigmatize individuals, has a terrible history of abuse, and "is potentially self-defining and sometimes embarrassing"—all reasons why everyone has a strong interest in not sharing it with individuals or companies.

Much has changed in the past two decades, but the basic privacy issues, especially those involving large collections of DNA samples, have yet to be solved, and the major suggested solutions all have serious drawbacks. Genomic privacy is still a work in progress. One approach is to "personalize" privacy, the way genomics seeks to personalize medicine. Each of us would be able to set our own limits on who could have access to our DNA samples and what uses they could make of them. As we'll explore later in this chapter, informed consent could be used to provide privacy level choices. Google thought about using this approach for its users but ultimately decided not to. Google's idea was to let their users decide for themselves where on a sliding scale they wanted to protect their privacy: minimal, medium, or maximum. We especially like the names they gave to people who opted for each of the three settings: kitten, cat, and tiger. Our view is that you should be a tiger when it comes to protecting your genomic information. It's your genome, as a matter of both property and privacy, and we think the more you know about it—and the uses others may make of it to the detriment of you and your family—the more you will be likely to want to protect it from others.

Before reading further, ask yourself if want your physician to have access to your whole-genome sequence in your medical record (this may become more likely as the cost of sequencing decreases and the quality of electronic health records increases). If the answer is yes (and a yes answer could make sense, even to a privacy tiger), think about who else you would be willing to share your genome with—for example, your spouse, your children, your parents. Then think about your employer, your coworkers, your life, health, and disability insurance companies, your credit card company, and your bank. Now we're in kitten territory. Finally, think about the government, including the police, the FBI, and the NSA. In all these cases, would you agree to have your genome linked to your medical record? What about your Facebook account or your OkCupid profile?

Does the perception that we may be moving slowly toward universal health insurance coverage in the United States make it reasonable for the government to maintain everyone's whole-genome sequence in a national data bank for planning purposes and perhaps for medical research? What about collecting DNA for identification to solve crimes and prevent terrorism? It may seem strange, but more effort has gone into creating DNA data banks for criminal and terrorism investigations than for medical research. That's why we begin our exploration of DNA data banks with the most recent U.S. Supreme Court case on DNA privacy.

Criminal DNA Data Banks

If you were stopped for running a red light, would you expect the police officer to take a DNA sample from you? What if the police officer arrested you on the spot? Should the police be able to take a DNA sample from you at the station without your consent? And once collected, should your DNA sample be permanently stored in a criminal database? Alonzo King was arrested and charged with assault for threatening a group of people with a shotgun. As part of routine booking in Maryland, a DNA sample was taken from him using a buccal swab. His DNA profile was compared to others in the FBI's national DNA data bank, the Combined DNA Index System (CODIS). The sample matched an unidentified DNA profile taken from a rape victim in an unsolved crime. King was convicted of the rape on the basis of the DNA match. He appealed, arguing that it was a violation of his Fourth Amendment rights to take a DNA sample from him before he was convicted of any crime.

All fifty states require collection of DNA from convicted felons. Twenty-eight states have laws that authorize DNA collection from arrestees. Justice Anthony Kennedy, writing for the majority of the U.S. Supreme Court, decided that taking DNA at the time of arrest was reasonable under the Fourth Amendment, primarily because the government's interest was "legitimate" (providing "a safe and accurate way to process and identify persons in custody"), and that "individual suspicion" is not necessary for processing. Kennedy characterized the DNA profile as the "fingerprint of the 21st century," although he did concede that the case would present "additional privacy concerns" if the police searched the DNA sample for medical information.

Justice Antonin Scalia wrote a biting dissent. He noted that the only reason to run King's DNA sample in the CODIS data bank was to try to solve a previous crime (a reason not permitted by the Fourth Amendment), not to identify King. Whether you agree with Justice Scalia or not (we do), he makes an unassailable observation about the "genetic panopticon" created by the decision: an arrest for any charge can now result in your DNA being collected and put in the national DNA data bank. Further, since DNA can lawfully be collected after conviction, the only arrestees for whom this Supreme Court decision will matter at all are those who have been acquitted of the crime for which they were arrested.

In balancing personal privacy with public safety, safety won out by a large measure (it almost always does, because safety seems immediate and vital, whereas privacy is less pressing and more abstract), even with no evidence to support it. With more than 10 million samples in criminal DNA data banks in the United States, our country already has the largest DNA data bank in the world. The United States also has the dubious distinction of having more of its citizens behind bars than any other country in the world. Is the Supreme Court right in concluding that DNA is just like a fingerprint, only better? A case from the European Court of Human Rights challenges this assumption.

The United Kingdom has been a world leader in collecting and using DNA profiles for criminal investigations since its first DNA dragnet, recounted vividly in Joseph Wambaugh's 1989 book, *The Blooding*. Alex Jeffreys's then new DNA profiling technique was used to conduct a DNA dragnet using blood samples from more than 5,000 men who lived in an area where two teenage girls had been brutally raped and murdered. Such DNA dragnets are unusual, but in 2014 Italian police took samples from more than 22,000 people to try to find the murderer of a thirteen-year-old girl. More routinely, DNA samples from arrestees are stored in data banks. The practice of police collecting DNA for suspect identification was initially justified to identify rapists and child molesters—groups the public has little sympathy for and whose privacy seems to come at the expense of victimized women and children. The use of DNA samples for identification has since expanded gradually but relentlessly to include more and more categories of criminal suspects.

The European case was brought by S, a minor who had been charged with attempted robbery when he was eleven and later acquitted, and Michael Marper, an adult who had been charged with harassment of his partner. The couple reconciled before trial, and the case against Marper was dismissed. S and Marper both asked that their DNA samples and profiles be destroyed on the basis that retention violated their right to privacy because the DNA samples contained "full genetic information about a person including genetic information about his or her relatives." The UK argued that its indefinite retention of DNA samples was of "inestimable value in the fight against crime and terrorism." The European Court of Human Rights agreed but was concerned that the UK made no distinctions in retaining DNA samples based on the gravity of the offense charged or the age of the suspect, that there were no time limits on retention, and that few opportunities existed to have the DNA samples and profiles destroyed. Similar to the concerns expressed by Justice Scalia in the *King* case, the European court was troubled that innocents were treated exactly the same as convicted criminals. The court accordingly instructed the UK to modify the procedures used in its DNA collection to permit the destruction of DNA samples and profiles from those not convicted of any crime and from minors.

In 2014 another European Court, the Court of Justice of the European Union, ruled that Europeans have a legal "right to be forgotten." The case did not specifically deal with genetics, but was much broader, giving individuals the right to require that large data holders, such as Google and Facebook, break links to personal information that was out of date, embarrassing, or just plain wrong. This real opinion provided a good counter to the fictional totally transparent society envisioned by Dave Eggers in *The Circle*. Eggers imagines a privacy graveyard where people are expected to share their entire lives with everyone else online. His Internet society (based on private industry) has three slogans, mirroring the three slogans of the dictatorship of *1984:* "Secrets are Lies," "Sharing is Caring," and "Privacy is Theft."

The U.S. Supreme Court demonstrated, in the 2014 case *Riley v. California,* that it understands how cell phones and the Internet have radically changed the nature of privacy. The police sought to access the information on an arrestee's cell phone on the basis that searching the phone was just like searching an arrestee's pockets, wallet, or purse. The Supreme Court was unimpressed, noting, "That is like saying a ride on horseback is materially indistinguishable from a flight to the moon. Both are ways of getting from point A to point B, but little else justifies lumping them together." Chief Justice John Roberts argued that the primary characteristic of a cell phone is its "immense storage capacity." This implicates privacy in terms of the different kinds of information it holds and its ability to convey far more than was previously possible, dating back to the purchase of the phone. Finally, in the words of the chief justice, it is the element of "pervasiveness" that characterizes information in "the digital age"—its sheer quantity. But the quality of the new digital data also matters: a browsing history found on a phone "could reveal an individual's private interests or concerns—

perhaps a search for certain symptoms of disease, coupled with frequent visits to WebMD." The Court did not specifically mention DNA data, but it would certainly find digitalized DNA data out of bounds in a police cell phone search, at least in the absence of a warrant.

Are Europeans more concerned about genetic privacy than Americans because of their experiences during World War II and the Cold War? It seems likely. It is also perhaps not surprising that one of the most powerful images of government oppression using information collection comes from the former Soviet Union. Aleksandr Solzhenitsyn, in his novel *Cancer Ward,* writes that in a totalitarian state people are obliged to answer questions on a variety of forms, and each answer "becomes a little thread" permanently connecting him to the local government center: "There are thus hundreds of little threads radiating from every man. . . . They are not visible, they are not material, but every man is constantly aware of their existence. . . . Each man, permanently aware of his own invisible threads, naturally develops a respect for the people who manipulate the threads . . . and for these people's authority." DNA implicates privacy even more than answers on forms—such as tax and census forms—because our DNA identifies us directly and can be seen as an integral part of who we (and our family members) are.

Mainstream scientists agree that it is critical to protect privacy in this context. Writing in support of the European Court's UK opinion, under the headline "Watching Big Brother," for example, *Nature* editorialized that the decision was especially timely, coming as it did just before the sixtieth anniversary of the Universal Declaration of Human Rights (UDHR): "The idea that the identity of a human can be revealed from [DNA] samples of any cell in his or her body is a symbol of the fact that every person is unique. The declaration of human rights [UDHR] asks us to treasure and honor all these unique individuals with respect for their autonomy—not to simply look for better ways to barcode them." The point could be made a different way. The two great "codes" of the twentieth century—the UDHR and the genome—should be seen as mutually reinforcing human dignity rather than as an opportunity to use one (the genome) to subvert the other (human rights).

Commercial DNA Data Banks

The steep drop in the cost of whole-genome sequencing has led to new proposals to develop national and international DNA research data banks filled with whole-genome sequences (and links to medical records and other health-related data). In the words of historian of science Hallam Stevens, who opened this chapter, genomics already is "a science obsessed with data," and many of its practitioners believe that "with enough sequence data and powerful enough computers . . . it will be possible to answer almost any question in biology, including and especially big questions about whole organisms, bodies and diseases."

The first real attempt to construct a large DNA database to explore diseases was in Iceland. The fact that it is an island with a small population (300,000) and a hundred years of a national health system with complete records made the prospect of combining DNA samples with medical records almost irresistible. A private biotechnology firm, deCODE Genetics, was formed to develop and link three data banks (one consisting of computerized medical records, a second of DNA samples, and a third of genealogical data) in an effort to locate disease-related genes. The project has been controversial since 1998, when Iceland's legislature approved the creation of a computerized Health Sector Database for deCODE's use in research. Consent has been a recurring problem for the company and Iceland's citizens.

George debated the company's founder and CEO, Kári Stefánsson, both at a national medical meeting in 1999 and in the *New England Journal of Medicine* in 2000. George argued for requiring individual informed consent, while Stefánsson argued instead for "community consent" via the Iceland legislature. The legal rules regarding privacy of the database took center stage again in 2009, when deCODE put the database up for sale to avoid bankruptcy. Ultimately the company was purchased by a much larger company (Amgen), and the data bank has remained in Iceland.

In 2015 deCODE announced that it had sequenced 10,000 genomes from Icelanders (2,600 of which were reported on in *Nature Genetics*). Combining the sequences with medical records and genealogical data, Stefansson said he could identify everyone in Iceland who had the *BRCA2* gene "at the push of a button," adding, it is "a crime not to approach these people" to let them know. But if there is any "crime," it is in analyzing an identifiable person's genome without consent. It was the first genomic databank to do this, but it is unlikely to be the last. We need better rules now. We think individuals have a right to access this information (if they want it), even though they did not authorize its creation. We also think individuals have a right to order the genomic information about them destroyed. The company is the custodian of the identifiable information, not the owner. Disclosure and destruction rules should, of course, be agreed to before a person gives permission for genome sequencing, and it is quite astonishing that deCODE did not have such an agreement with its depositors given its checkered history with informed consent.

The company had originally relied on "presumed consent" to create their database of medical records, but the Iceland Supreme Court later ruled that explicit informed consent was required. The ruling came in a case brought by the daughter of the late Guomundur Igolfsson, who died in 1991. His daughter, then fifteen years old, asked that information from her father's medical records not be transferred to the national Health Sector Database. Her request was denied, and she appealed to the Iceland Supreme Court. She argued that she had a personal interest in preventing the transfer of her father's records to the database, because "it is possible to infer, from the data, information relating to her father's hereditary characteristics which could also apply to her." The court ruled that Iceland's constitution requires the legislature "to ensure to the furthest extent that the information cannot be traced to specific individuals," and that existing procedures did not meet this standard. Therefore, the law that created the Health Sector Database was unconstitutional. This ruling killed whatever was left of deCODE's original project to create a countrywide computerized database of medical records, and the project was able to continue only after it adopted an explicit consent model.

These European cases support an individualistic view of privacy. They also indicate that the concept of genomic privacy is broad enough to protect the family unit from unwanted and unwarranted intrusion by both government and private actors. DNA is a family matter because, as we have mentioned elsewhere in this book, DNA provides information not only about the person from whom the sample is taken but also about the person's siblings, parents, and children.

China has much bigger ambitions. Its Beijing Genomics Institute, known as BGI, is the world's largest genomic research center, and it plans to create a genome bank with a million human genomes. When asked about privacy concerns, especially the role genomics could play in China's reproductive policy, BGI's president told *New Yorker* writer Michael Specter, "I don't care. Emperors have been ruling us for thousands of years. I know the government is watching us at all times. So what? I don't care about my personal privacy. It just doesn't matter." That attitude is, of course, the problem, not any kind of a solution.

The UK Biobank, perhaps the closest to the one President Obama envisions for the United States, aims to enlist 500,000 British citizens ages forty to sixty-nine to provide not only DNA samples but also medical records and answers to 250 personal questions, all for the cause of medical research. The goals of the project are laudable, but the consent rules should be fair and easily understandable. Current rules, which require research subjects to "relinquish all rights to their blood and urine samples, and give permission for access to their medical records at any time, even after death," seem extreme and unfair without a clear provision to discontinue "participation" at any time, including—as in Iceland—the right of surviving children to discontinue use of their parent's DNA samples. De-identification is also no substitute for consent, as it is becoming evident that DNA can be linked to identifying information in many cases. This is one reason why geneticist George Church asks people who want to have their genomes made part of his personal genome project (PGP) to waive their privacy rights. He now has 3,500 volunteers, with a goal of 100,000 and an ambition of a million, matching the Chinese project's goal. He also seems inclined to follow the Chinese view of privacy, although with the knowledge and consent of the volunteers. In his words, "What I really wanted was for everybody to have their genome and ideally everybody to share their genome."

Consent to DNA Data Banking

As anyone with an Internet connection already knows, individual consent is under broad attack in the United States, where it has historically been strongest. When you want to use almost any service on the Internet, you are asked to "agree" to the service's privacy policy. This policy is usually more than fifty pages long, and even if you wanted to read it, is incomprehensible to most people—although we get the bottom line: the company can do whatever it wants with the information it gets about you from your visit to their site. This, and related models such as "broad consent," "tiered consent," or "opt-out consent," have no place in the genomics realm, where others can use your genomic information against you in a variety of ways. Whenever the adjective *informed* is replaced by another adjective before the word *consent,* you should be suspicious that an effort is being made to reduce your role in deciding how your personal information will be used. Use of your information by others to discriminate against you is just one reason why you may want to control access to your genomic information. Another is that we think the ultimate battle to retain or modify medical information privacy will be won or lost in the battle over your control of genomic information. Until the enactment of the federal statute known as GINA (the Genetic Information Nondiscrimination Act), employers and health insurers could discriminate against you based on your genomic information. Others still can.

Antidiscrimination laws are necessary to protect people from genetic discrimination, but as we suggested in 1995 (in a proposed federal statute we drafted for the Ethical, Legal and Social Implications portion of the Human Genome Project, which we called the "Genetic Privacy Act"), privacy protection involves more than protection after genetic information has been obtained and shared with others. Effective privacy protection would also require personal authorization for at least four prior steps: collection of your DNA sample (for instance, from a blood sample or cheek swab), analysis of your DNA sample, storage of your DNA sample, and storage of the results from your DNA analysis. Of course, only after collection does the question of authorized use come up. So when you see large information bundlers, such as Microsoft, argue that we shouldn't worry about the collection of our data but only about the use of the data, you will understand that this is simply wrong. The best way to protect against misuse of our data is to prevent collection in the first place.

With the advent of whole-genome sequencing, your entire DNA sample could be digitalized, and there would be no need (except arguably for quality control) for data collectors to retain the DNA sample itself. Plans to combine large digitalized DNA data banks are well under way, and "investigators are clamoring for unified informed-consent documents that will allow them to compile genetic information into databases without creating a legal thicket of differing privacy protections." This strategy seems to be putting form (and forms) over substance, but a "unified" document could make sense if it contained sensible protections for individuals. In this regard, an editorial in *Nature* on UK Biobank got it right: it is a fundamental human right to determine how personal medical data are used, and exceptions to specific informed consent cannot be taken for granted. " Informed consent is not an obstacle to be overcome but a principle to be respected and cherished."

Research is necessary for medical progress, and for progress to be attained in genomics, research will require massive DNA data banks, "very big data," to search for correlations among genes, environment, family histories, the biome, and other variables that can affect your health. Big genome will quickly join big data in developing larger and more integrated DNA data banks, probably better termed "genome banks," for research. We encourage this development, because, as we have explored in a variety of contexts in this book, the science of genomics is much, much more complicated than virtually anyone recognized as recently as a decade ago, and it includes recognition of the critical roles of environment, the microbiome, and epigenetic effects. Progress in understanding genomics will require massive data banks of genomic and personal health information. But big data alone won't solve scientific problems. Data is just data; it's the interpretation that matters. And interpretation is much more difficult than collection.

Correlations may help us create hypotheses to test, but in medicine and public health, it is causal relationships that matter. When computers were new, the common lament was "garbage in, garbage out." The refrain now is "Big Error can plague Big Data."

Google Flu Trends shows, for example, some of the limitations of big data. Google believes that by monitoring all Google searches for terms related to the flu, it can apply algorithms to determine the extent of a flu outbreak and where it is spreading, and do it "several days" faster than the surveillance methods of the Centers for Disease Control and Prevention (CDC), which rely on reports from the field. The algorithms have been used in twenty-nine countries since the program was first launched in 2008. In 2013 Google Flu greatly overestimated how many people in the United States would get the flu at its peak. Some thought the overestimates were based on many healthy Googlers who were curious about how the flu season was going. Others, however, just thought that the CDC's methods remain better than the big data approach because, as the National Association of Healthcare Access Management put it, "The CDC model can control more factors than the Google model," including looking for respiratory viruses that are not the flu. Google is changing its algorithms, but it will be a while, if ever, before its big data approach is ready to replace traditional public health reporting and follow-up. Of course, sometimes neither approach will work, as in the universal failure to predict the Ebola epidemic of 2014.

Similar big data approaches have been used to direct cholera prevention projects in Haiti. The hope is someday to combine various data sources "to predict an outbreak before it begins." That's the hope. The reality is, as a proponent observer put it, that "information will only get us so far." In Haiti, for example, "equally pernicious forces, such as politics, nationalism, strained resources, and fear can, in a crisis, override even the best data tools." We're not against the movement toward big data in genomics, but it is very easy to oversell and overpromise, and you should feel no obligation to donate your DNA to a big data bank or a big data project. In making your decision, it may help to think of your DNA as your property. How would emphasizing your property rights instead of privacy rights affect your relationship with a genomic data bank? Put another way, does it make sense to redefine the relationship between a genetic data bank and a depositor from a researcher-subject relationship to a recipient-gift-giver relationship? The story of John Moore helps us answer this question.

In 1976, John Moore, a worker on the Alaskan pipeline, was treated for hairy cell leukemia by hematologist-oncologist David W. Golde at the University of California, Los Angeles (UCLA). As is standard treatment, Moore's spleen, which had enlarged from about eight ounces to more than fourteen pounds, was removed. Moore improved quickly. In an act that recalls the Henrietta Lacks case, Golde took a sample from the spleen and isolated and cultured an immortal cell line capable of producing a variety of valuable products, including one that stimulates the bone marrow to produce more white blood cells to fight infection in patients undergoing chemotherapy. In 1983, UCLA was granted a patent on the cell line. (figure 9.1) Moore said he never would have known about the cell line had Golde not called him in 1983 and told him he had "miss-signed the consent form" (circling I "do not" grant instead of I "do" grant UCLA all rights in "any cell line"). Moore sued UCLA and Golde for stealing his cells. The trial court ruled that Moore

United States Patent
Golde et al.

UNIQUE T-LYMPHOCYTE LINE AND PRODUCTS DERIVED THEREFROM

Inventors:	**David W. Golde; Shirley G. Quan, both of Los Angeles, Calif.**
Assignee:	**The Regents of the University of California, Berkeley, Calif.**
Appl. No.:	**456,177**
Filed:	**Jan. 6, 1983**

9.1 Patent of Mo cell line. David W. Golde, "Unique T-Lymphocyte Line and Products Derived Therefrom," U.S. Patent 4, 438, 032, filed Jan. 6, 1983, and issued March 19, 1985.

had no rights to his own cells, but an appeals court ruled that if Moore could prove theft of his cells, he could prevail.

The California Supreme Court ruled that Moore's physician should have disclosed his financial interest in using his patient's spleen for research and commerce. The court also ruled that Moore had no property interest in his cells after they had been removed from his body. In short, everyone in the world could own John Moore's cells except John! The court's reasoning was straightforward but unsophisticated. The court noted, for example, that California statutes governing the disposition of excised tissue, such as organ transplant laws, do not give property rights in organs to patients, but it failed to note that these laws only cover organs removed from corpses, who obviously can't own anything.

The real reason for the California ruling is the court's belief that "the extension of [property] law into this area will hinder research by restricting access to the necessary raw materials [and] destroy the economic incentive to conduct important medical research." Just as in the *King* case, where the U.S. Supreme Court easily concluded that an individual's interest in genetic privacy must give way to the state's interest in safety, the California Supreme Court concluded that Moore's property interest in his cells must give way to the interests of society (and private corporations) in (genomic) research and commerce.

In a later Florida case involving the collection of tissue samples and medical information from families affected by Canavan disease, a court ruled that under Florida law "the property right in blood and tissue samples . . . evaporates once the sample is voluntarily given to a third party." As in *Moore,* the judge worried that a finding for the plaintiffs would "cripple medical research." And in a case from Missouri involving prostate cancer samples from patients who had signed research consent forms that authorized the storage and use of their tissue for future research, a court ruled that the patients had transferred all their rights in the tissue to the researchers.

The courtroom conflicts in these cases were the result of real confusion over who had control of the tissue samples from which various products and genetic information were derived and what form of consent or authorization, if any, was required from the patients who supplied the samples. Traditional rules for clinical research don't work, because collecting and storing human tissue is not a research activity performed on humans, nor is it new, experimental, or even controversial.

The continuing controversies over the appropriate use of human tissue, including its use as a commercial product, suggest that human tissue donation needs its own rules and standards. Agreement on such standards has eluded human tissue collectors to date, but the Missouri prostate cancer case can help to move us to the next phase: regulating tissue donation and banking by creating, via statute or best practices, a formalized process for tissue donation. Think about whether you would donate a saliva or blood sample to a DNA data bank that would use the genome derived from it for medical research. What would you expect to be told about your "gift"? And would you let the data bank link to your electronic health record?

We believe a formalized process for human tissue donation (including donation for the purpose of genome derivation) should require explicit recognition that a gift is being made, identify the recipient of the gift, and specify whether the gift is conditional or unconditional. If a conditional gift is intended, the donor must specify the conditions, including any requirement that the tissue be destroyed if those conditions are not met. For example, "I want my DNA sample used for any medical research EXCEPT that involving mental illness or dementia." The statutory approach would be to draft a law similar to the Uniform Anatomical Gift Act, which governs the donation of dead bodies and tissue removed from corpses. Such a law would (or at least should) also determine under what circumstances, if any, people would be permitted to sell their tissue, and what buyers could do with it. Current federal law, for example, prohibits the purchase or sale of human organs for transplantation, and current national voluntary guidelines prohibit the purchase and sale of human ova, at least for stem cell research. Clear rules for collecting and storing human tissue would benefit both collectors and providers.

The Presidential Commission for the Study of Bioethical Issues took a stab at the privacy question in its 2012 report, *Privacy and Progress in Genome Sequencing*. Among its most important recommendations were that everyone involved in whole-genome sequencing adopt "clear policies defining acceptable access to and permissible uses of whole genome sequence data," and that federal and state governments adopt policies that "protect individual privacy by prohibiting unauthorized whole genome sequencing without the consent of the individual from whom the sample came." The commission also insisted on a "robust and workable consent process" that includes a description of whole-genome sequencing, how the data will be analyzed, stored, and shared, and the types of results from research or testing that individuals will get.

One major problem is determining what research use can be made of DNA samples already collected and stored. For example, would a committee or review panel made up primarily of donors be acceptable in granting consent for research use if it is transparent to the donors and representative of them? One suggested analogy is letting a friend or a group of friends order dinner for you in the event you are late. But delegating a dinner order to someone is quite different from delegating decision making about research. This is because you have no obligation to eat what has been ordered (and can likely get something else to eat even if you refuse what has been ordered), whereas the research will be done whether you like it or not. More than a simple delegation of authority is at stake here. The National Institutes of Health (NIH) recognized this in the context of reaching a 2013 agreement about future research conducted on the famous HeLa cells taken from Henrietta Lacks in 1951 without consent (discussed in chapter 4). The cells are now available around the world, but the ability to do whole-genome sequencing on them is new.

Following publication of a sequence of the HeLa cells by a German research team, surviving family members made it clear that they felt their privacy was being invaded without their authorization. As Jeri Lacks-Whye, Henrietta's granddaughter, put it (in words similar to the daughter in the Iceland case): "I think it's private information. . . . I look at it as though these are my grandmother's medical records that are just out there for the world to see." Working with Francis Collins, the family agreed that NIH would set up a committee. The committee, on which members of the Lacks family now sit, must approve any NIH-funded research that makes use of the Lacks data stored in the an NIH database, and publications that use the data must recognize Henrietta Lacks and her surviving relatives. No money would be paid to the Lacks family. Although NIH said this agreement did not create a precedent, it is hard to understand why not, or to understand why all contributors to DNA databases should not have an opportunity to vote on how their DNA should be used. A lot has changed since 1951, including the ability to keep in contact with tissue donors by cell phone and personal computers. As Jeri Lacks-Whye advised, "Have them involved, that's not only for HeLa sequences, but anybody who participates in research."

Gene Patenting and Big Data

The DNA "donors" in both the Moore cell line and Florida Canavan cases objected to patents on cells or DNA sequences that permitted others to profit from their cells and control their use. While they did not generalize this concern, others have, and cries of "no patents on life" have been heard for decades. Only in 2013 did the U.S. Supreme Court finally rule on the question of whether human genes are patentable (a form of ownership that gives the patent holder a twenty-year monopoly over use of his "invention"), deciding by a vote of 9–0 that they are not. The case can be dated from 1980, the year the U.S. Supreme Court decided, on a 5–4 vote, that a modified bacterium to which scientists had added four plasmids (which enabled the bacterium to break down various components of crude oil) was patentable. The patent claim was for "a non-naturally occurring manufacture or composition of matter—a product of human ingenuity 'having a distinctive name, character and use.'" The Court agreed that the *Chakrabarty* bacterium was new, "with markedly different characteristics from any found in nature," due to the added plasmids and the resultant "capacity for degrading oil." It did not matter to the Court that the methods employed to create the new bacterium had been in routine lab use for more than fifteen years, and the addition of the plasmids involved no new technology, such as use of restriction enzymes or in vitro recombinant methods.

After a research race largely financed by the National Institutes of Health, a private company patented two genetic sequences, now known as the "breast cancer genes," *BRCA1* and *BRCA2* (see chapter 8). Myriad Genetics developed a testing service, charging approximately $2,500 for testing a DNA sample to determine the cancer-predisposing mutations in these two genes. Myriad also charged anyone else who might offer genetic testing for these two genes with patent infringement. In 2007, author Michael Crichton, shortly after he published *Next,* his novel based largely on the *Moore* case, wrote in an op-ed in the *New York Times* that gene patents could cost lives because they are used "to halt research, prevent medical testing and keep vital information from you and your doctor." Moreover, in the case of breast cancer, they raise costs "exorbitantly"—from $1,000 to $3,000. Crichton concluded that the patenting rules left patients with no choice: "Don't like it? Too bad."

Shortly thereafter the American Civil Liberties Union and others brought suit against Myriad Genetics to have the patents on breast cancer genes declared illegal. George was at the Supreme Court for the oral arguments in early 2013. Leading genomic scientists were also in the courtroom, including James Watson, the Nobel laureate who codiscovered the double-helix structure of DNA, and Eric Lander, president of the Broad Institute of Harvard and MIT, who we have quoted previously. It was a big case, perhaps the biggest science-related case the Court had ever decided, and everyone knew it. The U.S. Patent Office wanted all Myriad patents upheld; the ACLU wanted both the patents on the breast cancer genes themselves and the patents on those same genes in which DNA had been synthetically modified in the laboratory by removing everything except its exon (coding) sequences—called complementary DNA (cDNA)—declared unpatentable. The Obama administration took an intermediary position: that the genes should not be patent eligible, but the cDNA ("synthetic DNA") should be.

The crucial legal issue (are genes patentable subject matter?) would be settled by the answer to a question of fact: did isolated DNA molecules, specifically the *BRCA1* and *BRAC2* genes, occur in nature in this form? Eric Lander had submitted a brief on this point, arguing that they did. He had read the lower court decision and decided that the judges were misinformed on the science. He cited twenty-four scientific papers in his brief to support this factual proposition. He further argued that because of an incorrect assumption (that is, incorrectly assuming that isolated DNA did not occur in nature), the lower court had resorted to unhelpful and unnecessary analogies, including whether a chromosome was like a leaf plucked from a tree, or like a kidney surgically removed from a human body, or even like a baseball bat carved from a tree trunk. Justice Stephen Breyer asked Myriad's lawyer whether he (Justice Breyer) correctly understood the Lander brief to say that isolated cells precisely identical to the *BRCA1* gene exist in nature: "Now, have I misread what the scientists told us, or are you saying the scientists are wrong?" Breyer ultimately was not persuaded that the lawyer for Myriad understood the science of genetics better than the scientists themselves.

Many observers were surprised at how little support any of the Justices showed for patenting whole genes and gene sequences, but after the oral argument, almost no one was surprised by the decision. Written by Justice Clarence Thomas for a unanimous Court, it ruled the way the Obama administration had suggested: genes that occur in nature are not patent eligible; genes that have been modified in the laboratory and do not occur in nature are. We found Justice Thomas's description of DNA and the distinction between isolated DNA segments and cDNA so well put, we've included it in Appendix A. We also liked this formulation: "It is undisputed that Myriad did not create or alter any of the genetic information encoded in the *BRCA1* and *BRCA2* genes. The location and order of the nucleotides existed in nature before Myriad found them."

The decision opened the way to whole-genome screening tests by any company that wants to do them or market them without any worry that the company might have to obtain a license from the estimated 3,000 companies and individuals who had claimed patents on naturally occurring genetic sequences or genes. Without this reassurance, few companies would likely pursue whole-genome sequencing as a business model.

Comments

The key paradox of the current state of whole-genome sequencing is summarized by the President's Commission for the Study of Bioethical Issues: "The majority of the benefits anticipated from whole genome sequencing research will accrue to society, while associated risks [that is, privacy risks] fall to the individuals sharing their data." The more we learn from DNA data bank research, especially about how to interpret the human genome, the more important privacy becomes. This is true in all its aspects: from secrecy, to protect ourselves and our families from discrimination; to anonymity, to permit us to volunteer our DNA to be used in medical research and in large data banks without fear of being identified or having the research findings used against us; and autonomy, to enable us to use genetic information about ourselves and our families as we see fit. Far from killing privacy, because of all the information our DNA contains, genomics can be said to have given privacy a new life.

The most important conclusion from this chapter is that your DNA is yours—it's your property, and it contains information that is most important to you—since it is unique to you. This gives you the legal and moral authority to decide what to do, and what not to do, with it. This includes the right to donate your DNA to a DNA research data bank (and to set limits on what it can be used for), as well as to decide not to make such a donation. Big data proponents argue that we should regulate data only at the stage of use and disclosure, not at the stage of collection. Thus, for example, they think it is fine for Google or the NSA to collect all the data it can—as long as they restrict how they use it. We don't. Once your DNA has been collected and stored, control over it becomes much more difficult and problematic. You should have the right not to have your DNA collected in the first place without your informed consent or authorization. Of course, even after collection, it should not be analyzed without your authorization as well.

The right to authorize analysis of your DNA also comes with a related right not to have your DNA analyzed or linked to your medical record. This is sometimes referred to as a "right not to know," or a right to be ignorant, but we think it is more accurately seen as freedom to live your life as you see fit. You have a right to live your life without being subjected to information that could make your life worse. This is, for example, why James Watson could agree to have his entire genome sequenced and made public but refuse to grant permission to search for genes related to Alzheimer disease—a condition he greatly feared. We all have the same rights—although not the same knowledge of genetics—as Watson.

In deciding whether to participate in a DNA research data bank—and it will literally take millions of individuals to make meaningful progress in genetic diagnosis and treatments—you should insist that your privacy be protected in ways you want and can monitor. No research should be done on your identifiable DNA sample (although even unlinked DNA can be identifiable) without your prior approval, and no information should be released to you or anyone else about your personal results without your prior agreement. All genomic research projects done with the banked DNA samples should be made public and the protocols approved by a board that includes significant numbers of DNA depositors. Results should be made public as soon as they are reasonably verified.

Genomic research using large-scale multinational DNA data banks is critical to applying genomics to clinical medicine. The public is rightly supportive of it—but people are more likely to donate their DNA and medical records to public, not-for-profit research data banks than to for-profit commercial data banks. This makes sense as the former is more likely to take your privacy seriously. Nonetheless, no matter what type of bank you decide to give the gift of your DNA, you should make clear your conditions on the gift, including what the data bank can and cannot do immediately with your DNA, and what uses of your DNA will require the bank to recontact you for authorization.

The next and final chapter is devoted to scientific speculation, which is common in the popular press and even scientific journals, but which may strike you as more science fiction than science fact. Where can, and where should, the new genomics take us?

WHEN THINKING ABOUT DNA DATA BANKS AND PRIVACY, CONSIDER THESE THOUGHTS

Privacy is a public good; your DNA is not.

Genomic privacy protects both our physical DNA and the information derived from our DNA and stored digitally.

Genomic privacy is a family matter because your DNA also informs on your family.

No one should collect, store, analyze, or use your DNA, or information derived from your DNA, without your permission.

DNA data banks should always be required to protect privacy and ensure informed consent, including guaranteeing you the right to withdraw your DNA and information.

CHAPTER 10

Genomics Future

Do not try to live forever. You will not succeed.

—George Bernard Shaw, preface to *The Doctor's Dilemma* (1911)

We have been primarily concerned with the ways in which our evolving uses of genomics in clinical medicine might reshape the practice of medicine in America. In our final chapter, we examine genomics future, and broaden our horizons to consider how the evolving genomics could affect both the broader society and even the world. This is speculative genomics. We ask you to think about the novel ideas of contemporary genomic researchers. Are they reasonable predictions of the future, or do you find that they are more like science fiction scenarios, created as precautionary tales to discourage experiments that might endanger our future as a species? We will also refer to science fiction to inform our discussion. As prominent genomics and synthetic biology researcher Craig Venter has observed, many of the greatest ideas in genomics "have been anticipated by myth, legend, and, of course, science fiction."

Our century can be viewed as one that attempts to redeem science from the destruction of physics (gray science)—as displayed in the atomic bomb and the doctrine of mutually assured destruction (MAD)—by redesigning nature through genomics (green science) to make life better for humans. For good or evil, as scientist and futurist Freeman Dyson has put it, "The 20th century was the century of physics and the 21st century will be the century of biology."

Most genomic futures feature improvements in human life and health. But at least some experiments along the way are dangerous. We may, for example, use genomics to try to change the nature of human beings—changes that could simultaneously cause us to change our notions of human rights and even of crimes against humanity.

There is already a brief history of promising but potentially dangerous genomic experiments. In the early 1970s, when the technique of recombining genetic sequences (recombinant DNA) was just being introduced, the primary experimental bacteria was *E. coli,* a common inhabitant of the human gut. Leading scientists met at Asilomar, a conference site in California, and agreed that no further recombinant DNA experiments using *E. coli* should be conducted until the safety of the procedure could be assured. Safety was ultimately achieved by restricting recombinant DNA experiments to laboratories that had elaborate methods to contain any dangerous bacteria that might be developed. In addition, a biological containment mechanism was added: the creation of a form of *E. coli* incapable of survival outside the laboratory. No similar moratoria were called by scientists until 2013, when a temporary moratorium (we think it should have been permanent) was called on experiments designed to see if a potentially dangerous H5N1 flu could be modified to make it communicable in mammals (ferrets). Both these experiments were designed to learn more about basic biology.

A set of proposed future genomic experiments is designed to go beyond bacteria and mammals, to use genomics to make a "better" human by modifying either the genes of an embryo or the brains or bodies of adults. In this chapter we examine some of the most prominent of these proposals—from both the scientific literature and contemporary science fiction. These include the merging of humans and machines to create immortals, the manufacture of "mirror humans" who are immune to disease, and the creation of posthumans or transhumans. How can we, as citizens of our global society, take part in defining the goals and methods of genomic research?

Life, Death, and Immortality

We all have thoughts about how the world will end, even though few of us obsess over them. Some favorites of ours include our sun expanding to incinerate us, climate change making Earth uninhabitable, a meteor striking Earth and destroying it, a plague of zombies, or a biological plague—like the flu or even Ebola—that destroys all human life. The role of genomics in these possible futures is most prominent in a plague scenario. In her wonderfully destructive apocalyptic trilogy (*MaddAddam*, *Oryx and Crake*, and *The Year of the Flood*), Margaret Atwood invents a world that has reverted (or evolved) into another Eden. All but a handful of humans have been destroyed by a biotoxin created by a disturbed young genetic engineer, Crake. A new species of humans, who will inherit what remains of Earth, the Crakers, have had all their destructive tendencies genetically engineered out, including racism, hierarchical thinking, territoriality, and sexual desire. They are "perfectly adjusted to their environment."

We'll come back to the Crakers and their new Eden. We begin with them to emphasize that at the margin it is often difficult to tell the difference between reasonable scientific speculation and science fiction. Moreover, in examining the speculations of some contemporary genomic scientists, we will not only discover where they think we might be going but will also be able to explore the ethical and legal implications of their visions. Atwood is not sanguine about the ability of scientists and corporate leaders to have insight into the implications of their work for humanity. In her words, "The people in the chaos cannot learn. They cannot understand what they are doing to the sea and the sky and the plants and the animals. They cannot understand that they are killing them, and that they will end by killing themselves."

Atwood's envisioned future may be (let's hope) overly pessimistic. But maybe not. Some scientists, acknowledging the vast destruction humans have already inflicted on the planet, have called for a "resurrection" of extinct species using genomics and cloning. Their goal is to transform the Jurassic Park of science fiction into science reality by using cloning technology to bring back extinct species such as the carrier pigeon and the woolly mammoth. The idea, variously referred to as the "Lazarus Project," "revival biology," and "the de-extinction movement," does seem more science fantasy than science fact. Nonetheless, serious genomic scientists, including Harvard Medical School's George Church, have endorsed it. At its best, Ryan Phelan of the "Revive & Restore" project argues that attempting to bring back extinct animals is a compelling way to call the world's attention to the threat that more animal species will soon die out. On the other hand, if the public believes that whenever a species becomes extinct, genomic science can bring it back through cloning, the urgency of species death may dissipate. Two other possibilities are that, like Crake, genomic engineers may wish not only to revive but to improve the now extinct species. Crake, for example, "improved" pigs (which he called Pigoons) by adding human brain tissue and organs suitable for human transplant.

Church has thought about how the woolly mammoth project might proceed and has described it in *Jurassic Park* terms. The process of using historical DNA and combining it with the DNA of an existing creature to try to recreate the historical creature is best illustrated in a cartoon in the middle of the movie *Jurassic Park* which explains how DNA can be taken from a modern frog and used to replace missing DNA from a long-dead dinosaur. Church would not follow this procedure, but instead he says he would begin with the genome of an elephant and attempt to modify it so the next-generation elephant would grow more hair. This is not a resurrection project but rather a genetic modification project. Craig Venter seems correct in dismissing the entire enterprise of reviving extinct species as "fanciful." This is because only a small number of DNA segments from the woolly mammoth have survived and they are degraded and highly fragmented.

The woolly mammoth "resurrection" project was left to perhaps our most famous living artist, Damien Hirst. Hirst procured the skeleton of a woolly mammoth, deconstructed it, gilded its bones with gold leaf, put it back together again, and encased it in a display box (figure 10.1). Of course, his creature is not alive. Hirst was not imitating nature but rather commenting on nature in a way that could tell us more about ourselves than science. Hirst observed that the mammoth "came from a time and place that we cannot ever fully understand [and] despite

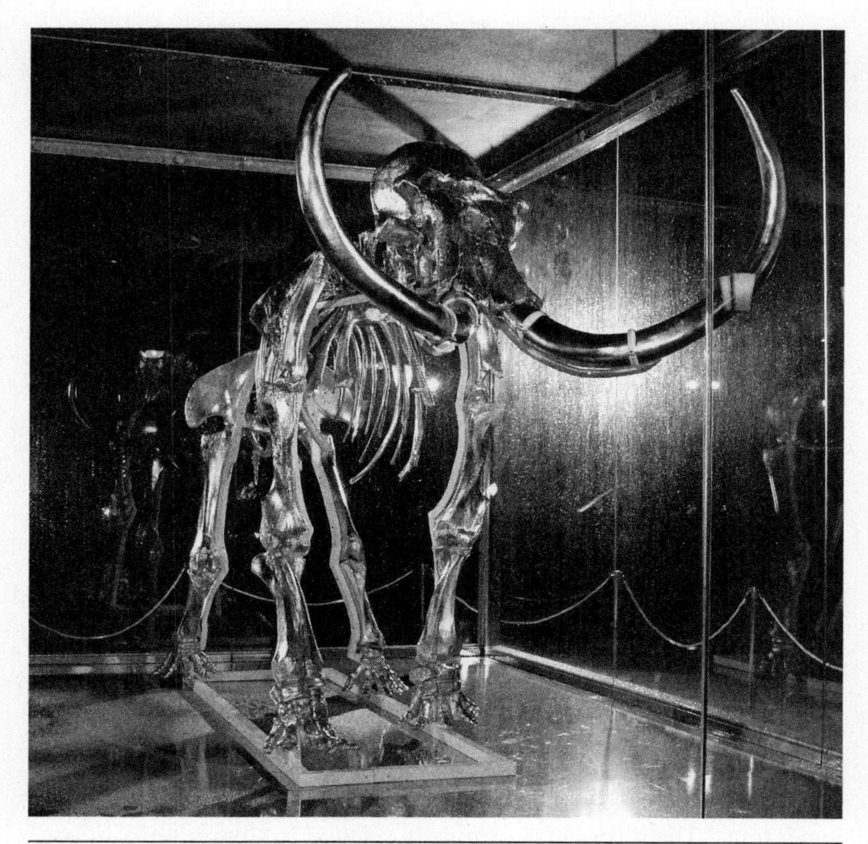

10.1 Damien Hirst's "Gone but Not Forgotten" (2014). Dave Bennett, "amfAR's 21st Cinema Against AIDS Gala Presented by Worldview, Bold Films, and BVLGARI – After Party," *WireImage*, May 22, 2014.

its scientific reality has attained an almost mythical status." Gold leaf was added to "combine science, history, and legend." Hirst concluded, "I've pitched everything I can against death to create something more hopeful." The woolly mammoth is "Gone but not Forgotten."

Hirst won the race to resurrect the woolly mammoth. No matter. Church has more ideas for using genomics to resurrect the past. Church has said, for example, that he thinks we will witness the birth of a Neanderthal baby in his lifetime. He has also speculated that cloning many Neanderthals would be good for human diversity and could even save the human race from extinction. We should not try to resurrect just a single Neanderthal, but should, he said, "create a cohort . . . [by cloning, who could] maybe create a new neo-Neanderthal culture and become a political force." This outrageous proposal caused little comment.

Public controversy followed instead from one sentence in Church's book *Regenesis*: "Supposing, then, that we have recreated the physical genome of Neanderthal man in a stem cell, the next step would be to place it inside a human (or chimpanzee) embryo, and then implant that cell into the uterus of an extraordinarily adventurous human female—or alternatively into the uterus of a chimpanzee." Church did concede that this could only happen "if human cloning becomes safe and is widely used." Press outlets treated an interview about his book as a serious attempt to find the "adventurous female." Headlines included, "Harvard Professor Seeks Mother for Cloned Cave Baby." Church credibly denied planning to do any such thing, but he did announce a few days after the interview that "hundreds and hundreds" of women "have volunteered to be surrogates [mothers] for the Neanderthal child."

To the best of our knowledge, Church is the first geneticist to suggest that we use genomics to reverse-engineer humans—to go back in time to resurrect our ancient and extinct human relatives, like Neanderthal man. London's *Daily Mail* put Church's hypothetical project in context: "The phrase 'Frankenstein science' has never been so apt." Pressed to explain his enthusiasm for this type of genomics experiment, Church conceded that we may not actually want to recreate the entire Neanderthal, but just introduce Neanderthal genes into the DNA of modern *Homo sapiens*—the DNA that codes for brain or "neuronal pathways, skull size, a few key things." He continued, "That could give us what we want in terms of neural diversity. I doubt that we are going to particularly care about their facial morphology."

Picking genes to create a "better baby" is the dream of many a genetic engineer. It still remains in the realm of science fiction. Will it ever become reality? And what characteristics would make a "better baby"? George Church again helps us begin our exploration. As he told *Der Spiegel*, "The de-extinction of a Neanderthal would require human cloning." The interviewer responded that human cloning "is banned." Church replied (correctly), "That may be true in Germany, but it's not banned all over the world. And laws can change."

Whether genomic scientists ever succeed in producing a human clone, the delayed genetic twin of an existing human (more than fifteen years after the birth of Dolly the sheep, it has yet to happen), we can learn a lot from the international movement to ban human cloning. And the most important things will be about life, not science, about values, not technique. The reason political leaders around the world called for a ban on applying the Dolly-the-sheep cloning technique to humans, if not always well articulated, is that replication of a human by cloning could radically alter the very definition of a human being by producing the world's first and only human with a single genetic parent. Cloning a human can also be viewed as uniquely disturbing because it is the manufacture of a person made to order, it represents the potential loss of individuality, and it symbolizes the scientist's unrestrained quest for mastery over nature for the sake of knowledge, power, and possibly profit.

Fiction has been far more influential than science writing in producing society's mixture of fascination and horror regarding cloning, as exemplified by films such as *Blade Runner*, *Sleeper*, and *Jurassic Park*. Science fiction suggests that governments, corporations, wealthy individuals, and rogue scientists have multiple motives to clone. A type of hypercloning was, for example, the basis for governing in Aldous Huxley's *Brave New World* (1932). The key to social control in Huxley's society was the "Bokanovsky Process," in which a single embryo is stimulated to divide into ninety-six identical copies. These ninety-six embryos were then artificially gestated together under identical conditions designed to produce four basic classes of workers: Gammas, Deltas, Epsilons, and Alphas. Specific "batches" were conditioned to perform socially useful tasks and to love performing them. Their happiness with their lot in life was reinforced chemically: the drug soma kept them contented. Huxley's combination of genetic engineering of bodies and chemical engineering of minds seems as unlikely today as it did seventy-five years ago. That is because there are many more efficient ways of creating large numbers of troops or terrorists. Physical and psychological conditioning can turn teenagers into soldiers in a matter of months, rather than waiting some eighteen to twenty years for clones to grow up and be trained. Cloning has no real military or paramilitary uses. The profit motive is much more powerful today. As the author of *Jurassic Park,* Michael Crichton, put it: "The commercialization of molecular biology is the most stunning ethical event in the history of science." Science writer Barry Werth has also noted that while science needs "facts, data, evidence, rigor," business can be founded on profile building using "blue smoke and mirrors."

Science fiction can help us identify the major ethical and social policy issues raised by cloning a human, but it cannot tell us how to proceed. There are four basic regulatory models to choose from: the market, professional standards, government regulation, or an outright ban. In the medical care and medical research context, we have consistently urged medical specialty organizations to set and follow practice and ethics standards. Unfortunately, to date the relevant professional medical and genomics associations have not been able to move beyond Crichton's commercialized market-consumer model. As we noted more than two decades ago, the existing practice in reproductive genetics is to provide consumer-patients with whatever they want (and can pay for), rather than to develop a professional model that sets meaningful practice and ethics standards or takes seriously the welfare of resulting children. If anything has changed over the past two decades, it is probably the emergence of the market as an even stronger force than professional self-identity in shaping professional standards, as well as the simultaneous commodification of sperm, eggs, embryos, and pregnancy—and arguably, human children.

A treaty to ban human cloning was widely debated but ultimately rejected at the United Nations. A substitute, nonbinding, "Declaration on Human Cloning" was adopted by the UN General Assembly in 2005. The declaration calls on all member states to prohibit cloning to make a baby and prohibit genetic engineering (provisions we favor). But it goes further, prohibiting human cloning to make medicine—a provision insisted on by the United States but one that we (and thirty-four countries that voted against the declaration) think is misguided and unnecessarily restricts important scientific research. The absence of a treaty means that the world reverted to its default mode, which we have termed "ethical arbitrage." This means simply that researchers and corporations are free to take their genomic experiments, including cloning experiments, to the country with the most lax regulatory scheme. The cloning treaty failed primarily because the United States insisted that any cloning treaty reflect its position that all human cloning, including using embryos to make stem cells, should be outlawed. The U.S. debate on cloning was significantly deformed by its relationship to the abortion debate and the religious beliefs of many that a human embryo should be treated like a baby from the moment of its creation. It was this belief that led President George W. Bush to limit federal funding for stem cell research using cells derived from human embryos to those that had been created before August 9, 2001, the day he made his funding decision. President Obama continued this funding ban. That means that virtually all stem cell research since 2001 has been privately funded or funded by individual states, and this has fostered a "Wild West" mentality in which experiments are driven more by hope of profits than pursuit of science. As we wrote in *Nature* during the 2004 presidential campaign, it also led the President's Council on Bioethics to concentrate on articulating an ethical justification for the funding ban, which it was unable to do.

There were other objections to the cloning treaty. One important one was the belief held by proponents of human modification that neither cloning nor germline modifications should be considered so inherently dangerous that they should be outlawed. George had suggested that cloning and germline modifications should be thought of as "species-altering procedures." The objection to this category of procedures was that it is too vague and too inclusive. This critique has some merit, and it is reasonable to think that only species-altering procedures that are "species endangering" should be outlawed. That is because it may be possible to genetically alter the human species, or members of the human species, in ways that do not put the species itself at substantial risk. For example, changing the skin color of all humans to green would alter the species but would not put humans at risk of either extinction or transformation into another species.

Species-Endangering Experiments

Technology has made it possible for some humans to put all humans at risk of extinction. Kurt Vonnegut, a survivor of the firebombing of Dresden, provides a good starting point for understanding species-endangering experiments in his profound 1963 novel *Cat's Cradle*. *Cat's Cradle* is narrated by Jonah, who is writing his own book about what prominent Americans were doing on the day the atomic bomb, a product of gray technology, was dropped on Hiroshima. Collecting information for his book, he interviews Dr. Breed, a scientist and associate of one of the inventors of the atomic bomb (Dr. Hoenikker), who insists that he only does "pure science," not military science. He concedes he was once asked by a Marines general to make something that would eliminate mud, because the Marines were sick of fighting in mud. Breed never developed this product. Hoenikker did invent—but never experimented with—an investigational mud-destroying product, called "ice-nine." When ice-nine is ultimately field tested, it crystallizes not just the water in mud around it. In a chain reaction, it crystallizes all water, ending life on earth.

The probability of actually creating a new weapon or life form that would threaten the very existence of the human species is low, but it seems reasonable to conclude any quantifiable probability of human extinction must be taken seriously. For example, if cloning is a necessary precondition to genetically modifying humans in a way that endangers the future of the species as species, the international community could reasonably decide to include it in the species-endangering category. On the other hand, using cloning (the creation of human embryos by somatic cell nuclear transfer for use in research) to make medicine (by harvesting stem cells from them) would fit into neither category, since this type of cloning neither alters nor endangers the human species. Quite the opposite: it offers humanity the prospect of medical benefit.

As Vonnegut suggests in *Cat's Cradle*, the most familiar global annihilation scenario is not a genetically driven experiment at all, but a large-scale nuclear exchange. Other candidates for species-endangering experiments include the introduction of inheritable genetic alterations, the development of human-machine cyborgs, the direct creation of new human pathogens, the enhancement of existing human pathogens, the weaponization of an untreatable biological agent or toxin, and the creation of new weather patterns. More specific illustrations include the development of a new and more lethal strain of smallpox as a bioweapon and the use of inheritable genetic alterations to create "superior" humans. The subject matter is technologically driven and recognizes that new technologies have made it possible for some humans to put all humans, the entire species, at credible risk of extinction. Of course, only prevention matters in this extreme sphere: an extinct or radically altered species cannot prosecute its destroyer, or even convene a truth and reconciliation panel. As Margaret Atwood's end-of-the-world trilogy underlines, only unsuccessful attempts to destroy the human species are prosecutable or forgivable.

As these examples suggest, the human species is most likely to be endangered in two ways. The first is through an existing or new weapon of mass destruction—biological, nuclear, or chemical—that could destroy all or most members of the human species. Even a one-in-a-million chance that the human species would be destroyed is reason to prohibit scientists from developing a weapon or a pathogen (it may be a chance worth taking to develop a lifesaving drug, however, or a cure for cancer). Use of these dangerous military and technological creations can be seen as a subset of genocide, and like genocide, the most important strategy is to prevent potentially genocidal weapons from being developed in the first place. Once ice-nine was developed, its use was inevitable.

The second species-endangering category, and the one we focus on because it directly involves genomics, is a modification or alteration in some members of the human species that could lead to the demise of the species qua species by direct replacement with a new species, or by initiating what George has termed "genetic genocide" (in which the modified humans destroy the unmodified humans, seeing them as subhuman, or the unmodified humans destroy the modified humans, seeing them as a threat to their continued existence). But is preserving the human species such an important goal that we should forgo possible improvements in the species through genomics? Francis Fukuyama probably articulated this concern in the clearest language when he noted, "It is impossible to talk about human rights—and therefore justice, politics, and morality more generally—without having some concept of what human beings actually are like as a species."

Posthumans and Transhumans

Jaron Lanier, who has been described as a "megawizard in futurist circles," is skeptical of our ability to predict the future (on Earth, let alone in the universe) but nonetheless takes the posthuman movement seriously. In *Who Owns the Future?* he quotes from notes taken at the Singularity University (located next to Google in Mountain View, California): "Your mind is software. Program it. Your body is a shell. Change it. Death is a disease. Cure it. Extinction is approaching. Fight it." Lanier's own bottom line about his fellow futurists: "What most outsiders have failed to grasp is that the rise of power of 'net based monopolies' [like Google and Facebook] coincides with the new sort of religion based on becoming immortal."

Posthumanists despair of our species entirely, looking forward enthusiastically to a future in which we will happily leave *Homo sapiens* (and our genes) behind and merge with machines (this merger—of *Homo sapiens* with machines—is sometimes referred to as the "Singularity") to become immortal. At this point, we will simply leave our current human species behind; *Homo sapiens* will look like what a species of monkeys looks like to us today. In one vision, we will exist only as digital information. In another it is when humans and computers merge to create a new life form.

In a debate with a posthuman enthusiast at Yale a few years ago, George irritated his opponent so much that the opponent declared, "I don't care about the human species." For us (and I would suspect an overwhelming majority of the members of the human species), that's the problem, or at least a problem with posthumanism's prophets. It's one thing for our species to disappear or merge with machines as an inevitability; it's quite another for some humans (or a single human, like Crake) to make this choice for all of us.

It is not suicide, but immortality that is on the minds of most posthumanists. This seems to be the dream of Google—whether achieved by genetics or big data. Its project was announced in a cover article in *Time* magazine: "Can Google Solve Death?" The authors of the article (and likely many Americans) seem to believe that if anyone could solve death, Google could. The basis for their belief is digitalization: "Medicine is well on its way to becoming an information science" (true) and "doctors and researchers are now able to harvest and mine massive quantities of data from patients" (not yet). Their bottom line: "Google is very, very good with large data sets." Probably true but unlikely to have anything to do directly with immortality. But for us, and we're sure for readers of chapter 8 on cancer, it was Google cofounder Larry Page who articulated an otherworldly view of the capacity of big data to extend our lives. He said he had just learned that even if we could prevent or cure all cancers, we would only be able to add about three years to average life expectancy. From that he concluded that "solving cancer [is not] this huge thing that'll totally change the world." He continued, "There are many, many tragic cases of cancer, and it's very, very sad, but in the aggregate, *it's not as big an advance as you might think*" (emphasis added).

We take a different view: we think curing cancer would be a giant deal to billions of people around the world (especially those who will develop and die of cancer), and although immortality is impossible (George Bernard Shaw was right when he advised, "Do not try to live forever. You will not succeed"), there is no hope of major life extension without solving the cancer problem. As "individualistic" Americans we are much more interested what happens to us that in what happens to the "average person." Within a year of the Google death cover, for example, *Time* ran a special health double issue with a baby on the cover with the caption, "This baby could live to be 142 years old." You should ask yourself whether you find that prospect comforting or horrific.

Damien Hirst's diamond-encrusted skull strikes us as a fitting symbol both of Google's project and America's inability to come to terms with human mortality (figure 10.2). Hirst titled the skull "For the love of God." It is based on the reaction of his mother to it ("For the love of God Damien, what will you do next?") The American answer: We will conquer death (or decorate our lifeless skulls to symbolize our striving for immortality).

A year after it announced plans to challenge mortality, Google announced another of its "moonshot" projects, a study of what goes on inside the human body, called Baseline Study. As reported in the *Wall Street Journal,* the study will collect not only the entire genome of thousands of volunteers but also their medical histories, information on how they metabolize food, nutrients, and drugs, their heart rate, respiration, and the content and actions of their microbiomes, and much more. They will also be monitored by Google-invented wearable devices to collect additional information. The goal is to identify new biomarkers that can be used to predict (and hopefully cure) diseases before they start. This project implicates all of the privacy and informed consent issues we highlighted in chapter 9, and not just in relationship to genomics.

10.2 Damien Hirst's "For the Love of God" (2007). *Wikimedia Images,* September 15, 2007.

In the Baseline Study project, Google seems to be taking data collection on humans to its logical extreme, viewing the human body as simply the site of a massive amount of data—data that can be collected and analyzed to explain life itself. Craig Venter has analogously suggested that at some point it will be possible to digitalize a life form, send the electronic version of the life form to another planet, and reconstruct the life form from its data. The model for this concept is *Star Trek*'s transporter. Of course, reconstructing such a complex life form as a human may never be possible. But constructing microbes from a genomic blueprint may be. As Venter himself suggests, "It doesn't require a great leap to think that, if Martian microbes are DNA-based, and if we can obtain genome sequences from microbes on Mars and beam them back to Earth, that we should be able to reconstruct the genome."

Venter goes on to suggest that this "synthetic version of the Martian genome" could be used to "re-create Martian life for detailed study" without having to confront the logistical nightmare of actually going to Mars and bringing back a biological sample. Nor is Venter's vision limited to our solar system: "If this process can work from Mars, then we will have a new means of exploring the universe and the hundreds of thousands of Earths and Super-Earths being discovered by the Kepler space observatory." We could, for example, send our own DNA instruction book into the universe with the hope that another civilization might be advanced enough to know how to re-create us from our genome.

The posthuman movement is also being taken seriously in contemporary fiction. Dan Brown's bestselling *Inferno,* for example, has a transhumanist genetic engineer, Bertrand Zobrist, as its somewhat sympathetic villain. Zobrist has developed a powerful genetic weapon, an airborne vector virus capable of modifying human DNA to induce sterility. His goal is to use the virus to make a third of humans on Earth sterile and prevent humans from overbreeding to an unsustainable population. The discussion among the characters as to whether countergenetic engineering techniques should be tried to reverse the effects of the sterility virus includes points for and against genetic engineering itself. Both arguments are most effectively presented by Sienna, a disciple of Zobrist. First, Sienna presents the argument against genetic engineering: "The human genome is an extremely delicate structure . . . a house of cards. The more adjustments we make, the greater the chances we mistakenly alter the wrong card and bring the entire thing crashing down. . . . Even if you designed something you thought might work, trying it would involve *reinfecting* the entire population with something new. . . . Zobrist's actions were *reckless* and extremely dangerous."

Sienna also makes the argument in favor of inducing massive genetic alterations: "The transhumanist movement is about to explode from the shadows. One of its fundamental tenets is that we as humans have a moral obligation to *participate* in our evolutionary process . . . to use our technologies to advance the species, to create better humans—healthier, stronger, with higher-functioning brains. . . . [G]enetic engineering is just another step in a long line of human advances."

Futurist Ray Kurzweil, who joined Google in 2013 to work on artificial intelligence, believes he will live long enough to live forever. He thinks the date for the Singularity is 2029, the year he believes advances in genomics and nanotechnology will be occurring so fast, under the law of accelerating returns, that we will merge with our machines and become immortal. Kurzweil envisions the creation of a digitalized brain (software) that we can copy and download into an infinite variety of bodies (hardware). To those who worry that supercomputers will replace and destroy humanity, Kurzweil's response is that attempted regulation of this technology would require a totalitarian-type government that would simply drive technological innovation underground, where it would be even more dangerous. Self-regulation (the market) is proclaimed as the answer. While all techno-optimists resist government oversight, some also reject the Kurzweil human-machine merger model of the future. Genetic optimist Juan Enriquez, for example, sees the real merger as genomic: a merger of gene therapy, epigenetics, and proteomics that will enable humans to chart their own evolution to become better humans. In his words, "Forget the singularity—biology will trump technology."

Consent is a necessary precondition for ethical human experimentation, but it is not a sufficient justification: more than consent is at issue in species-endangering experiments, including experiments to produce the posthuman. An individual may provide consent to be a subject in an experiment, and in some cases, parents may consent on behalf of their children. But no one has the right or moral authority to consent to an experiment that puts the entire human species at risk. That is because the harm risked in a species-endangering experiment is not to the individual (who may or may not be harmed, and could even be benefited) but potentially to the entire species. If the risk of radically altering or destroying the human species is too high (more than 1 percent?), then consent is not relevant. Similarly, consent of the victims is no defense for committing crimes against humanity, such as slavery or torture.

George Church would, for example, need the consent of his hypothetical "extremely adventurous female" to clone his Neanderthal as a necessary precondition of the experiment. But her consent, while necessary, would not be sufficient. We would also need some sort of "best interests" (imputed) consent from the Neanderthal. But most important, as Church himself concedes, the final arbiter of whether the experiment could be done should be "society." Church doesn't say how society could make the decision. We suggest that something like a species-wide institutional review board, or IRB, should be established on the global level, with authority to approve species-endangering experiments (where the actual risk of endangerment is very low) on behalf of the inhabitants of Earth. Yes, we realize this sounds like science fiction too.

In 2015 two groups independently called for a moratorium on using a new and powerful genome editing technology called CRISPR-Cas9 on human embryos (as well as sperm and eggs) to attempt to make a better baby. One group, led by the president of a biotech company, Edward Lanphier, simply wanted to call a halt to any research on modifying human embryos which they saw as both risky and nontherapeutic. In Lanphier's words, "We are humans, not transgenic rats." The other group wanted to continue research on the editing technology to address the safety issues and was not ready to condemn using the procedure to try to create a better baby. Their conclusion was that what is needed now is an "open discussion of the merits and risks of human genome modification by a broad cohort of scientists, clinicians, social scientists, the general public, and relevant public entities and interest groups."

Another contemporary real-world example that illustrates the potential utility of a public review board involves studies designed to create a new and more dangerous version of H5N1 (bird flu) influenza virus that can be efficiently transmissible between mammals. The laboratory animal of choice for this experiment was the ferret. Publication of the work was controversial, many worrying that the modified H5N1 influenza could be a blueprint for the creation of a bioterror weapon. Others argued that the work was too dangerous to be done in the first place, a view we share. This view gained credence when in 2014 it was disclosed that the NIH had discovered that smallpox samples which should have been destroyed decades ago had been left in a storage room unprotected, and the CDC had negligently let loose live anthrax samples that it believed it had killed, as well as a particularly dangerous strain of flu. All the director of the CDC could say was that "the culture of safety needs to improve." That's an understatement. Bioterrorism expert Thomas Inglesby noted that if these accidents were taking place in the United States, they were likely happening around the world. In dangerous research, he argued, "everyone has to start with the assumption that we have human systems that are fallible. There's no such thing as perfect systems."

We think the lack of a global review mechanism to examine species-endangering experiments means that no such experiments can lawfully or ethically be conducted today. This says nothing, however, about the future in which such a representative and accountable body, which could be well short of world government, could exist. A similar suggestion was made by astronomer Martin Rees in discussing the chances that CERN's (European Organization for Nuclear Research) Large Hadron Collider, the world's largest and most powerful particle accelerator, could, when in operation, create a small black hole that would destroy the planet (the odds were vanishingly low, and thankfully, it didn't). In his words, "No decision to go ahead with an experiment with a conceivable 'Doomsday downside' should be made unless the general public (or a representative group of them) is satisfied that the risk is below what they collectively regard as an acceptable threshold."

Fact or Fantasy?

Martin Rees adopts Vonnegut's ice-nine scenario as one possible outcome of the supercollider experiments. Scientists' consistent use of science fiction to illustrate potential risks of their own experiments has failed to persuade many nonscientists that fiction can be useful in making real-world science decisions. For example, human enhancement proponent Timothy McConnell has argued that our proposal to outlaw species-endangering experiments, at least until an accountable global review body can be formed, seems to be motivated by "fear" of genetic enhancement stoked by reading science fiction. He specifically mentions *Frankenstein, Dr. Jekyll and Mr. Hyde,* and *The Time Machine.* McConnell rejects the likelihood that genetically enhanced humans will kill the unenhanced, terming it a conclusion that demands justification. We, on the other hand, think it is a risk that only the species can authorize.

Genetics ethicist Eric Juengst also believes that the real issue is discrimination, noting that George himself had summed up this species-distinction problem with a new term, *genism*. We have two responses: human history provides sufficient justification for assuming that the powerful will dominate and subjugate the weak, even with strong antidiscrimination laws; and the real question is the burden of proof in this policy debate. Should the opponents of species-endangering experiments have to demonstrate that it is more likely than not to result in human extinction, or should (as we think) the precautionary principle be used to put the burden of proof on would-be experimenters to demonstrate (to our global review body) that their experiment will not result in species extinction?

Human extinction could, of course, be caused by outside forces we cannot control (for instance, a large meteor hitting earth, climate change, or even an alien attack). Of greater concern to us are the actions we can control or prevent. As we suggested at the beginning of this chapter, Margaret Atwood provides a relevant cautionary tale in *Oryx and Crake.* Crake, a gifted but disturbed young genetic engineer, graduates from creating genetically fused novel animals to creating a happiness pill, "BlyssPluss" (a pill not too different from Huxley's soma), finally creating a whole new subspecies of humans, the Crakers. The BlyssPluss pill is designed to protect humans against all sexually transmitted diseases, provide unlimited libido, and prolong youth. It becomes so popular that Crake can (and does) unleash a worldwide lethal pandemic by adding a toxin to it ("It was a rogue hemorrhagic [amazingly quickly causing bleeding from eyes and skin, convulsions, and death]. . . . [T]he bug appeared to be airborne."). As for Crake himself, except for his new humans, the "fear, the suffering, and the wholesale death" of the rest of humanity, the old humans, did not touch him. "Crake used to say that *Homo sapiens* was not hard-wired to individuate other people in numbers above two hundred, the size of the primal tribe."

Atwood's musings (as articulated by Crake) are not that different from the musings of geneticist George Church in *Regenesis*. As Atwood notes in her acknowledgments in *MaddAddam*, hers is a work of fiction, but she "does not include any technologies or biobeings that do not already exist, are not under construction, or are not possible in theory." When George Church decided to take his own ideas about reengineering the human species to the public, he chose as his co-author Ed Regis, the author of a book on "science slightly over the edge," *The Great Mambo Chicken and the Transhuman Condition*.

Church's concept of transhumanism is to create what he calls "mirror humans." These are transhumans designed by "changing the handedness of an entire organism and all of its components, so that you have a mirror image of everything from the macro level all the way down to the atomic level." Church's transhumans would look just like current humans but "would be radically different in terms of resistance to natural viruses and other pathogens." It would "almost be as if two separate species of humans existed simultaneously," although "mirror humans should have an unusual smell" and could not sexually reproduce with existing humans. Church agrees that his experiment to produce mirror humans would be risky. Interaction of mirror molecules with existing molecules is "unpredictable," and "careful screening of mirror molecules by computational methods or by actual experiments will be necessary to ensure safety." Church seems to think regulation of his experiments is not necessary or even possible because the science is unstoppable. Like others before him, he seems to believe that the only thing scientists can't control is themselves: "Regulations . . . can be circumvented by anyone who is sufficiently determined to evade them."

Church explains his ultimate goal in the most ambitious language of any we have seen: "The [human] genome should become not just the genome of one lonely being or one planet. It should become the genome of the Universe." This vision mirrors that of physicist Frank Tipler, who sees humanity reaching the "Omega Point": the "ultimate state of the universe" where life has "gone everywhere" and "becomes omnipotent . . . and omnipresent." This vision becomes reality only if humans have "evolved" into nanoparticles capable of travel at near light speed—or, in Craig Venter's vision, we roam the universe as electronic data in search of a civilization capable of reconstructing our bodies and brains from our genomic blueprints.

It is, of course, much easier to make pronouncements about what should not be done than about what we should do. It is, for example, easy to say that we should not destroy our planet or our species by genetically re-engineering ourselves or releasing genetically modified biotoxins into the environment. The harder question is what we should do to try to prevent other humans from doing either of these things, or even whether we should worry about these possible futures at all. We think it unexceptional to suggest that you should join with those who want to use genomics for the good of humanity and our health, and try to prevent genomics from being used for evil. We think, for example, that it is reasonable to promote the use of cloning technology for new medicines, including regenerative medicine, while simultaneously opposing the use of cloning to make babies who are genetic duplicates of an existing human or the use of genetic engineering techniques to make a "better" baby. We think it is also reasonable to oppose resurrection experiments to bring back the woolly mammoth and carrier pigeons as simply unscientific and a waste of time and resources. Likewise, the proposal to resurrect Neanderthal man seems completely misplaced and even dehumanizing. Likewise, we think it is reasonable to oppose the Singularity project, although you may think, with Kurzweil, that it is inevitable with or without our help.

The most difficult genomic issue is trying to genetically modify a human embryo to create a better baby. It is a truism that evolution has in its own slow way created better babies over the centuries. No one would seriously argue (at least we wouldn't) that humans have reached their limit and cannot be improved. On the other hand, it is certainly not clear that we have the knowledge or ability to determine what will be "better" for the next generation—at least beyond a few immunities, like immunity to cancer. Even here, however, there may be unknown side effects, such as early mental illness, that would make cancer immunity not worth having.

Concluding Thoughts

We decided to write this book because we believed that the evolving science of genomics has the potential to radically change both how we think about ourselves, and the type of health care system we will have in the near future. This change will include incorporating electronic health records complete with our sequenced genomes, new genomic analytics that can help predict future diseases, new ways to treat and possibly cure complex diseases, including cancer and diabetes, and new ways to determine the effect various drugs (and foods) are likely to have on you based on your genome. We think that the messages conveyed by our genomes are so massive in quantity, and so potentially beneficial in quality that they will cause physicians and others to want to modify or eliminate traditional medical ethics doctrines of informed consent and privacy. We think this would be a mistake, making medicine both more impersonal and less responsive to patients than it is today.

Genomics will be adopted by our health care system because it reinforces the last three of the four basic characteristics of American medicine: it is wasteful, technologically-driven, individualistic, and death-denying. There is no new medical technology as seductive and pervasive as genomic technology, a conclusion that can be gleaned simply from the mantra-like sayings from the genomics industry, including the magical "thousand dollar genome"; the mythical "right drug, for the right patient." We simply cannot resist new technology and will welcome genomics into our lives and clinics.

Individualistic is, of course, the essence of genomics—we will treat you (or at least your genome) as unique, since you are the only person on the planet who has this genome (the essence of "personalized medicine"). This is not just good advertising, that is what most Americans actually want and expect from their physicians. Finally, as especially this chapter has focused on, we are death-denying. On one level we know we are mortal; but in our day-to-day lives we strongly deny this through our actions, and most Americans support an aggressive research program that has immortality (or at least living for 142 years) as a reasonable goal, no matter how unreasonable it actually is. All this is to say that genomics, although still in its infancy in American health care, has the potential to hit the health care system hard, and to vastly increase its cost because of its direct impact on the four characteristics that already make it by far the most expensive health care system in the world.

This means that unless we can control the introduction of genomics into medicine in a way that retains basic American legal and ethical values of informed consent and privacy, and unless we can get some control over the pricing of new drugs, genomic medicine will have to be rationed, and will likely only be available to the wealthy and well-insured. An alternative future is one based on a modified Medicare-for-all health plan where everyone is entitled to basic health care (including most of the new genomics), but no one is entitled to medical interventions that are not cost-effective or that simply prolong dying.

On an individual level (and mostly this book is about you as an individual and family member) our hope is that we have provided you with enough information, together with pro and con arguments, so that you can make an informed decision about the uses you, your family, and your physicians will make of the new genomics. We also hope this book will help you make informed decisions about what uses of genomics you think should be surrounded with privacy and consent, and what uses of genomics should be regulated or even outlawed. If we have helped or encouraged you to think critically about these complex issues, we will take this as some measure of success in sending our own uncoded genomic messages.

WHEN THINKING OF GENOMIC FUTURES, CONSIDER THESE THOUGHTS

Science fiction often informs science and can inform us about personal and policy issues.

The way genome scientists and biotech entrepreneurs envision the future has much in common with science fiction.

We humans now have the capacity to put our entire species at risk of extinction.

Species-endangering experiments should be treated like a crime against humanity.

Immortality is not a characteristic of being human.

Cloning to make medicine is a reasonable scientific pursuit; cloning to make a baby is not.

There are many categories of proposed posthumans, most of which require that our genomes (and bodies) be replaced with electronic bits.

Before we try to make a "better baby," we should develop a global consensus on the characteristics that make human life unique and worthwhile.

Because it reinforces three major characteristics of American medicine as technologically-driven, individualistic, and death-denying, genomics will be eagerly embraced by American medicine, even if it leads to rationing.

Acknowledgments

This book draws widely and at times deeply on our interactions with our colleagues, as well as with medical, law, and public health students over the past three decades. We want to specifically acknowledge some individuals who have helped us in major ways. First, colleagues who have read all or major parts of the book and have given us thoughtful advice (which we have not always taken) on the text. These individuals include all of the full-time members of the department of Health Law, Bioethics, and Human Rights: Leonard Glantz, Michael Grodin, Wendy Mariner, and Patricia "Winnie" Roche. Sherman himself was a visiting scholar in the department in 2013. Other professional colleagues who offered their knowledge and criticism include Lynn Bush, Robert C. Green, Diana Bianchi, Nick Argy, and Brian Skotko. Family members with special expertise who were also kind enough to contribute to this project are Shelley Elias, Kevin Elias, and Mary Annas. We'd also like to acknowledge the support of Boston University, especially of the dean of the School of Public Health, Robert Meenan, and Boston University President Robert Brown. Strong support for the department's scholarship and advocacy has continued under the school's new dean, Sandro Galea.

The staff of the department of Health Law, Bioethics, and Human Rights was tireless and perceptive in supporting us in this project, especially Alicia Orta (who went on to purse a degree in genetic counseling), Jessica Walsh, and Gina Duong. We also greatly appreciated the help of our graphics consultant, Ann Blattner. And special thanks to our primary research assistant, Kathleen Joseph, who worked on this book with us over her sophomore, junior, and senior years in college, and who will begin medical school at Boston University in the fall.

We presented early versions of chapters 6 and 7 to the Health Law Professors Conference, and George presented earlier versions of chapter 10 at a human rights workshop in Amsterdam and to a health and human rights class at the University of Texas at Austin School of Law (where John Robertson was kind enough to provide engaging commentary).

We want to thank our agent, Elizabeth Evans, at Jean V. Naggar Literary Agency, and our editor at HarperOne, Gideon Weil. Finally, we should mention two people we have not met, but whose books had a major influence on *Genomic Messages*: James Gleick, author of *The Information,* and Daniel Kahneman, author of *Thinking, Fast and Slow.*

APPENDIX A

DNA and the Human Genome

Genes form the basis for hereditary traits in living organisms. . . . The human genome consists of approximately 22,000 genes packed into 23 pairs of chromosomes. Each gene is encoded as DNA, which takes the shape of the familiar "double helix" that Doctors James Watson and Francis Crick first described in 1953. Each "cross-bar" in the DNA helix consists of two chemically joined nucleotides. The possible nucleotides are adenine (A), thymine (T), cytosine (C), and guanine (G), each of which binds naturally with another nucleotide: A pairs with T; C pairs with G. The nucleotide crossbars are chemically connected to a sugar-phosphate backbone that forms the outside framework of the DNA helix. Sequences of DNA nucleotides contain the information necessary to create strings of amino acids, which in turn are used in the body to build proteins. Only some DNA nucleotides, however, code for amino acids; these nucleotides are known as "exons." Nucleotides that do not code for amino acids, in contrast, are known as "introns."

Creation of proteins from DNA involves two principal steps, known as transcription and translation. In transcription, the bonds between DNA nucleotides separate, and the DNA helix unwinds into two single strands. A single strand is used as a template to create a complementary ribonucleic acid (RNA) strand. The nucleotides on the DNA strand pair naturally with their counterparts, with the exception that RNA uses the nucleotide base uracil (U) instead of thymine (T). Transcription results in a single strand RNA molecule, known as pre-RNA, whose nucleotides form an inverse image of the DNA strand from which it was created. Pre-RNA still contains nucleotides corresponding to both the exons and introns in the DNA molecule. The pre-RNA is then naturally "spliced" by the physical removal of the introns. The resulting product is a strand of RNA that contains nucleotides corresponding only to the exons from the original DNA strand. The exons-only strand is known as messenger RNA (mRNA), which creates amino acids through translation. In translation, cellular structures known as ribosomes read each set of three nucleotides, known as codons, in the mRNA. Each codon either tells the ribosomes which of the 20 possible amino acids to synthesize or provides a stop signal that ends amino acid production.

DNA's informational sequences and the processes that create mRNA, amino acids, and proteins occur naturally within cells. Scientists can, however, extract DNA from cells using well-known laboratory methods. These methods allow scientists to isolate specific segments of DNA—for instance, a particular gene or part of a gene—which can then be further studied, manipulated, or used. It is also possible to create DNA synthetically through processes similarly well known in the field of genetics. One such method begins with an mRNA molecule and uses the natural bonding properties of nucleotides to create a new, synthetic DNA molecule. The result is the inverse of the mRNA's inverse image of the original DNA, with one important distinction: Because the natural creation of mRNA involves splicing that removes introns, the synthetic DNA created from mRNA also contains only the exon sequences. This synthetic DNA created in the laboratory from mRNA is known as complementary DNA (cDNA).

Changes in the genetic sequence are called mutations. Mutations can be as small as the alteration of a single nucleotide—a change affecting only one letter in the genetic code. Such small-scale changes can produce an entirely different amino acid or can end protein production altogether. Large changes, involving the deletion, rearrangement, or duplication of hundreds or even millions of nucleotides, can result in the elimination, misplacement, or duplication of entire genes. Some mutations are harmless, but others can cause disease or increase the risk of disease. As a result, the study of genetics can lead to valuable medical breakthroughs.

—Justice Clarence Thomas
for a unanimous U.S. Supreme Court,
Association for Molecular Pathology v. Myriad Genetics,
133 S.Ct. 2107 569 U.S. ______ (2013).

APPENDIX B

Limitations of Screening Tests

The purpose of a screening test is to determine if you have or don't have a certain condition or even a certain gene or mutation. In some cases the test may be inconclusive and may need to be repeated. However, screening results may not be accurate. The result of the test may indicate that you have a condition or disease you don't have, a "false positive." The test may also fail to detect a disease or condition that you do have, a "false negative." Usually there are additional tests that are conducted to confirm (or reject) the original finding.

The *sensitivity* of a test is how likely a test is to identify a true positive. The specificity of a test is the probability that a negative test result is a true negative. For example, if a screening test with 99 percent sensitivity has a positive result, there is a 99 percent chance that the person screened has the condition or disease. Unfortunately, the ability of a test to detect a disease or condition varies with the prevalence of the disease in the testing population. In a population where everyone has the condition we are looking for, the test will be accurate 99 percent of the time. But if only one in every thousand patients we are screening has the condition, even a test that has a 99 percent sensitivity and a 99 percent specificity, a patient with a positive result actually has only a 9 percent chance of having the disease! This 9 percent value is called the *positive predictive value* of the test, and it is calculated by dividing the number of true positive results of a test by the total number of positive tests. Extrapolating to rarer conditions affecting fewer than 1 in 100,000 creates even higher numbers of false positives and an even lower positive predictive value. Of course, if there are no people in the population you are screening who actually have the disease or condition, all positive results will be false positives.

Because the prevalence of a disease in a symptomatic patient population is much higher than the prevalence in an asymptomatic patient population, a positive result is much more likely to be a true positive in the symptomatic population. A good example is the current practice of breast cancer screening using mammograms. Imagine you are a health care practitioner. If you do a mammogram on a patient with a breast mass, it is more likely that an abnormal mammogram will correctly indicate a true positive rather than a false positive. This is because the prevalence of disease is much higher in a population of women with palpable masses. Likewise, because the prevalence of breast cancer in an asymptomatic population is much lower, an abnormal mammogram in this population is much more likely to be a false positive. (See the discussion of breast cancer in chapter 8 for more on this topic.) One solution to this dilemma is to personalize screening tests based on age and health status. Screening, some have suggested, should be offered to patients who are at high risk and thus are most likely to benefit from screening. Understanding the high false positive rates in screening tests should help you contextualize the news of a positive result. We caution you to be skeptical of screening results and urge you not to immediately assume you have a disease following a positive screening. It is important to proceed with diagnostic testing following any positive screening results.

Much genomic testing involves screening asymptomatic patients for rare conditions, and therefore is likely to result in a high number of false positive results. When genomic testing is performed, patients should not only be counseled on the myriad findings which are of unknown significance, but also that even positive results of known significance may not be real, especially if the condition is rare. The enormous complexity and sheer size of the human genome makes erroneous results likely. For example, 23andMe uses a microarray that reports 99.99 percent reproducibility for detecting SNP variants, meaning that they estimate the error is about 0.01 percent. While the error rate for a single SNP is low, 23andMe is not designed to examine a single SNP variable. Rather, it aims to examine thousands of SNPs in a single genome. It has been estimated that with an error rate of 0.01 percent applied to a million SNPs, at least a third of 23andMe customers have an error in the results of one of their "interesting" SNPs. Additionally, the genomic coverage, which measures the average number of times a base in the test genome is matched to a probe and is an indication of the accuracy of data, is variable. Most scientific publications require a coverage depth of 10 times to 30 times. In a recent study, the 23andMe microarray had an average coverage depth of 28.4 times, but 4.3 percent of bases were read at a very low coverage depth (less than 5 times).

As more and more genomic studies are conducted, researchers are revisiting previous studies and finding that much of what we thought we knew about the relationship between the genome and disease may not be true. A recent report on cancer genomics in *Nature,* for example, is titled, "Lists of cancer mutations awash with false positives," and describes how the use of improved models and reanalysis of cancer genome data decreased the number of genes possibly associated with cancer from 450 to 11. Another study, which evaluated over 600 positive associations between common gene variants and disease, found that only 166 associations had been studied more than 3 times. Of the 166 which had been studied repeatedly, only six were consistently replicated. As the science improves and more data is collected, potential pitfalls can multiply. This is not to suggest that we abandon the investigation but rather that we proceed with caution, recognizing the limitations present in the evolving field of genomics.

In the case of prenatal testing, when positive results can create the possible realization of parents' worst fears, a positive screening result must be tempered with the understanding that positive results for rare conditions are often inaccurate, and must be confirmed with a diagnostic test. At the 2013 American College of Medical Genetics and Genomics annual meeting presenters discussed the real problem of both false positive and false negative results in noninvasive prenatal testing. Cases of what were believed to be fetal aneuploidies (an abnormal number of chromosomes), actually reflected maternal tumor DNA. If confirmation testing had been bypassed, not only would perfectly healthy fetuses have been terminated, but maternal cancer would have proceeded undiagnosed.

Nate Silver, in his book *The Signal and the Noise,* notes statistical pitfalls that have direct applicability to genomics. He points out that statistical analysis of data frequently shows an association or correlation of data elements, but that statistical correlation is very different than causation. A simple example is high glucose levels in patients with diabetes. High sugar levels do not cause diabetes but are a manifestation of the disease. Another example is the higher incidence of automobile accidents in younger drivers. Does youth cause accidents or is it inexperience or a higher incidence of reckless behavior in younger drivers or both? There are likely other factors which account for the correlation as well. In genomics research certain genes are associated with a higher incidence of a particular disease, but this association does not mean causation. Many other factors such as epigenetics and the microbiome could be important. All of these reasons support comprehensive counseling prior to genomic screening.

Gilbert Welch goes even further in his book *Overdiagnosed,* arguing that since all of us harbor at least some genomic abnormalities, "we can all be shown to be at high risk for some disease." From this proposition he concludes: "So the new world of personal genetic testing has the potential to make all of us sick and arguably poses the greatest threat of overdiagnosis." Our hope is that if you have a better understanding of the limitations of much of genomic testing, you will be better able to use the results in ways that make your life better, not worse.

Notes

Chapter 1: The Coming Flood of Genomic Messages

1 **sending of a message**: James Gleick, *The Information* (New York: Pantheon Books, 2011), 296.

3 **into different kinds of relationships**: Hallam Stevens, *Life Out of Sequence: A Data-Driven History of Bioinformatics* (Chicago: University of Chicago Press, 2013), 8, 69.

5 **on metaphors for DNA**: Dorothy Nelkin and M. Susan Lindee, *The DNA Mystique: The Gene as a Cultural Icon* (Ann Arbor: University of Michigan, 2004); Ruth Hubbard and Elijah Wald, *Exploding the Gene Myth* (Boston: Beacon Press, 1993).

5 **Borges's Library of Babel**: Jorge Luis Borges, *Labyrinths* (New York: New Directions, 1962), 51–58.

5 **DNA is not linear**: E. Pennisi, Inching toward the 3D genome, *Nature* 347: 10 (2015)

6 **nonsense, synonym, and redundancy**: Gleick, *The Information*, 1, 299.

6 **regulate expression of distant genes**: J. R. Ecker et al., "Forum: ENCODE Explained," *Nature* 489 (2012): 52–55.

7 **amount of data**: E. Lander, quoted in G. Kolata, "Bits of Mystery DNA, Far from 'Junk,' Play Crucial Role," *New York Times*, September 5, 2012, A1.

7 **Lander commented that**: quoted in G. Kolata, Projects Sheds Light on What Drives Genes, *New York Times*, February 19, 2015, A16.

7 **despite the progress**: editorial, Beyond the genome, *Nature* 518: 273 (2015).

7 **known unknown**: Donald Rumsfeld, *Known and Unknown* (New York: Sentinel, 2011), xiii–xiv.

8 **$1,000,000 interpretation**: T. Caulfield, J. Evans, A. McGuire, et al., "Reflections on the Cost of 'Low-Cost' Whole Genome Sequencing: Framing the Health Policy Debate," *PLoS Biol. 11* (November 2013): e1001699.

9 **on MaddAddam trilogy**: Margaret Atwood, *Oryx and Crake* (New York: Anchor Books, 2004); *The Year of the Flood* (New York: Nan A. Talese/

Doubleday, 2009); *MaddAddam* (New York: Nan A. Talese/Doubleday, 2009).

9 **private as your diary**: G. J. Annas, "Privacy rules for DNA databanks," *JAMA* 270 (1993): 2346-50.

10 **to our aging selves**: W. Safire, "Sleazy Senate Inquiry," *New York Times*, October 25, 1993, 11A.

10 **a life's meaning**: A. Jureic, "Life Writing in the Genomic Age," *Lancet* 383 (2014): 776–77.

10 **shaped by the past**: Christine Kenneally, *The Invisible History of the Human Race* (New York: Viking, 2014), 312.

11 **Genetic McCarthyism**: R. C. Green and G. J. Annas, "The Genetic Privacy of Presidential Candidates," *New England Journal of Medicine* 359 (2008): 2192–93.

11 **mass-shooter gene**: G. Kolata, "Seeking Answers in Genome of Gunman," *New York Times*, December 24, 2012, D5.

11 **"thinking slow"**: Daniel Kahneman, *Thinking, Fast and Slow* (New York: Penguin), 2011.

13 **on Angelina Jolie Pitt**: A. Jolie, "My Medical Choice," *New York Times*, May 14, 2013, A25.

14 **you know about yourself**: "*23andMe TV Spot*" (television commercial), www.ispot.tv/ad/7qoF/23-and-me.

15 **on 23andMe and the FDA**: G. J. Annas and S. Elias, "23andMe and the FDA," *New England Journal of Medicine* 370 (2014): 985–88, and sources cited therein.

15 **The Patient Will See You Now**: Eric Topol, *The Patient Will See You Now* (New York: Basic Books), 2015.

15 **The FDA is in continuing talks**: A. Pollack, F.D.A. Reverses Course on 23andMe DNA Test in Move to Ease Restrictions, *New York Times*, February 19, 2015.

16 **My mother's ovarian cancer**: Angelina Jolie Pitt, "Diary of a Surgery," *New York Times*, March 24, 2015, A23; Pam Belluck, "Experts Back Actress in Choices for Cancer Prevention," New York Times, March 25, 2015, A3.

16 **on Sergey Brin**: T. Goetz, "Sergey Brin's Search for a Parkinson's Cure," *Wired Magazine*, June 22, 2010, www.wired.com/magazine/2010/06/ff_sergeys_search/all/.

16 **disease in the world**: "Parkinson's Disease," *23andMe*, https://www.23andme.com/pd/.

20 **Can you imagine?**: C. Y. Johnson, "Scientist Hoping to Ease Interpretation of the DNA 'Book of Life,'" *Boston Globe*, May 19, 2014. Green's genome also revealed "a few million variations, 109,000 of which could initially be considered medically relevant." Of those, 11,900 were identified by computational analysis to have an effect on a protein. Of these, 1,800 were common enough to warrant checking against data banks, after which 16 rare mutations (including the Treacher Collins variant) remained (as reported by David Cameron in *Harvard Medicine*).

20 **mirrors four major characteristics**: G. J. Annas, "Reframing the Debate on Healthcare Reform by Replacing our Metaphors," *New England Journal of Medicine* 332 (1995): 744-7, 1995.

Chapter 2: Personalized (Genomic) Medicine

23 **they are not arriving at the doorsteps of our patients**: C. Lenfant, "Clinical Research to Clinical Practice—Lost in Translation?" *New England Journal of Medicine* 349 (2003): 868–74.

24 **Reinventing American Health Care**: Ezekiel Emanuel, *Reinventing American Health Care* (New York: Public Affairs, 2014), 1–9.

25 **better individual treatment**: National Academy of Sciences, *Toward Precision Medicine: Building a Knowledge Network for Biomedical and a New Taxonomy of Disease* (Washington, DC: 2011).

25 **influence their health outcomes**: M. H. Chin, "How to Achieve Health Equity," *New England Journal of Medicine* 371 (2014): 2331–32.

26 **family history information**: S. Kaminer, "Sizing Up the Family Gene Pool," *New York Times*, February 24, 2012, MM13.

27 **My Family Health Portrait**: "My Family Health Portrait: A Tool from the Surgeon General," https://familyhistory.hhs.gov/FHH/html/index.html.

28 **by about 20 percent**: T. M. Frezzo, W. S. Rubinstein, D. Dunham, et al., "The Genetic Family History as a Risk Assessment Tool in Internal Medicine," *Genetic Medicine* 5 (2003): 84–91.

28 **predictions for cardiovascular disease**: N. P. Paynter, D. I. Chasman, G. Pare, et al., "Association between a Literature-Based Genetic Risk Score and Cardiovascular Events in Women," *JAMA* 303 (2010): 631–37.

29 **gallstones, [or] hemochromatosis**: A. Gutierrez, "Warning Letter Re: Personal Genome Service," *Food and Drug Administration*, www.fda.gov/ICECI?EnforcementActions/WarningLetter/2013/ucm376296.htm.

29 **manage what you can't**: Elias and Annas, "23andMe and the FDA."

29 **the FDA was correct**: Elias and Annas, "23andMe and the FDA."

29 **on Jim Mathis poll**: J. Schulte, C. S. Rothaus, J. N. Adler, et al., "Screening an Asymptomatic Person for Genetic Risk—Polling Results," *New England Journal of Medicine* 370 (2014): 2442–43.

29 **across the street**: "Ancestry," *23andMe*, https://www.23andme.com/ancestry/.

30 **family even mean?**: A. Jacobs, "Are You My Cousin?" *New York Times*, February 2, 2014, SR1.

30 **on Relman becoming a patient**: A. Relman, "On Breaking One's Neck," *New York Review of Books*, February 6, 2014, 61.

31 **on the Cheesecake Factory**: A. Gawande, "Big Med," *New Yorker*, August 13 and 20, 2012, 52–63.

32 **anyone who needs care**: A. Gawande, quoted in C. Nobel, "Attention Medical Shoppers: What Health Care Can Learn from Walmart and Amazon," *HBS Working Knowledge*, http://hbswk.hbs.edu/item/6658.html.

32 **ball bearing factory**: G. J. Annas, "Reframing the Debate on Health Care Reform by Replacing Our Metaphors," *New England Journal of Medicine* 332 (1995): 744–47.

33 **surgeons adopt checklists**: Atul Gawande, *The Checklist Manifesto* (New York: Picador), 2009.

33 **avatar physician**: R. Cook and E. Topol, "How Digital Medicine Will Soon Save Your Life," *Wall Street Journal*, February 21, 2014.

34 **place in the world:** "Church Family History Records Lead to Groundbreaking Genetic Research," *Newsroom: The Church of Jesus Christ of Latter-Day Saints*, July 18, 2008, http://www.mormonnewsroom.org/article/church-family-history-records-lead-to-groundbreaking-genetic-research.

35 **more than thirty states:** D. B. Kraybill, K. M. Johnson-Weiner, and S. M. Nolt, *The Amish* (Baltimore, MD: Johns Hopkins University Press, 2013), 4.

35 **on genetic research among the Amish:** V. A. McKusick, J. A. Hostetler, and J. A. Egeland, "Genetic Studies of the Amish," *Bulletin of Johns Hopkins Hospital* 121C (1964): 1–4; M. S. Lindee, "Provenance and the Pedigree: Victor McKusick's Fieldwork with the Old Order Amish," in Alan Goodman, Deborah Health, Susan Lindee, eds., *Genetic Nature/Culture* (Berkeley: University of California Press, 2003), pp. 41–57.

37 **on genetic research in Saudi Arabia:** B. McKay and E. Knickmeyer, "Saudi Researchers Mount Genome-Sequencing Push," *Wall Street Journal*, February 5, 2014, B8.

37 **biases the judgment:** Arthur Conan Doyle, *A Study in Scarlet and the Sign of the Four* (New York: Harper & Brothers, 1904), 28.

37 **clinical expertise and patient values:** D. L. Sackett, S. E. Straus, W. S. Richardson, and W. Rosenberg, *Evidence-Based Medicine: How to Practice and Teach EBM* (Edinburgh: Churchill Livingstone, 2000), 1.

38 **on PSA screening:** H. C. Cox, "Quality of Life Guidelines for PSA Screening," *New England Journal of Medicine* 367 (2012): 669–71.

39 **fruit and vegetable intake:** M. Ruffin, D. Nease, A. Sen, et al., "Effect of Preventive Messages Tailored to Family History on Health Behaviors: The Family Healthware Impact Trial," *Annals of Family Medicine* 9 (2011): 3–11.

41 **and laboratory testing:** D. McCormick, D. H. Bor, S. Woolhandler, et al., "Giving Office-Based Physicians Electronic Access to Patients' Prior Imaging and Lab Results Did Not Deter Ordering of Tests," *Health Affairs* 31 (2012): 488–96.

42 **act on the information:** D. B. Goldstein, "Growth of Genome Screening Needs Debate," *Nature* 476 (2011): 27–28.

Chapter 3: Nature, Nurture, and the Microbiome

45 **Abigail Zuger wrote that her own experience:** A. Zuger, "Genes Tell Only Part of Story," *New York Times*, February 17, 2015, D4.

45 **that it is genetic:** B. Marshall, quoted in M. Azad, "A Bold Experiment," *Nature* 514 (2014): S6–7.

48 **on 2012 Nobel Prize:** N. Wade, "Cloning and Stem Cell Work Earns Nobel," *New York Times*, October 9, 2012, A9.

48 **human embryonic stem cells:** Y. G. Chung, J. H. Eum, and J. E. Lee, "Human Somatic Cell Nuclear Transfer Using Adult Cells," *Cell Stem Cell* 14 (2014): 777–80; M. Yamada, B. Johannesson, and I. Sagi, "Human Oocyte Reprogram Adult Somatic Nuclei of a Type 1 Diabetic to Diploid Pluripotent Stem Cell," *Nature* 501 (2014): 533–36.

49 **on Woo Suk Hwang:** D. Cyranoski, "Cloning Comeback," *Nature* 505 (2014): 468–71.

49 **or an oak tree:** Ian Wilmut, Keith Campbell, and Colin Tudge, *The Second Creation: Dolly and the Age of Biological Control* (Cambridge, MA: Harvard University Press, 2001).

49 **on Mengele:** Gerald L. Posner and John Ware, *Mengele: The Complete Story* (New York: Cooper Square Press, 2000), 29.

50 **Francis Galton:** F. Galton, "The History of Twins as a Criterion of the Relative Powers of Nature and Nurture," *Journal of the Anthropological Institute of Great Britain and Ireland* 5 (1876): 391–406.

50 **twin study:** N. J. Roberts, J. T. Vogelstein, G. Parmigiani, et al., "The Predictive Capacity of Personal Genome Sequencing," *Science Translational Medicine* 4 (2012): 133ra58.

51 **prediction will remain probabilistic:** David Altshuler, quoted in G. Kolata, "Capacity of Genome to Predict Is Limited," *New York Times*, April 3, 2012, D5.

52 **prevention of NTDs:** "Spina Bifida and Anecephaly Before and After Folic Mandate—United States, 1995–1996 and 1999–2000," *Morbidity and Mortality Weekly Report* 53 (2004): 362–65.

53 **folate fortification of foods:** J. L. Mills and C. Signore, "Neural Tube Defect Rates Before and After Food Fortification with Folic Acid," *Birth Defects Research: Part A, Clinical and Molecular Teratology* 70 (2004): 844–45.

54 **McDonaldization of society:** George Ritzer, *The McDonaldization of Society* (London: Sage, 2013). See also T. Caulfield, "The Obesity Gene and the (Misplaced) Search for a Personalized Approach to Our Weight Gain Problems," *Wake Forest Journal of Law and Policy* 5 (2015): 125–145.

54 **two to nineteen years are obese:** "Overweight and Obesity," *Centers for Disease Control and Prevention*, http://www.cdc.gov/obesity/data/adult .html.

54 **on diabetes statistics:** "National Diabetes Statistics, 2011," *National Diabetes Information Clearinghouse*, http://diabetes.niddk.nih.gov/dm/ pubs/statistics/#fast.

55 **on MODY:** M. Shepherd, A. C. Sparkes, and A. T. Hattersley, "Genetic Testing in Maturity Onset Diabetes of the Young (MODY): A New Challenge for the Diabetic Clinic," *Practical Diabetes* 18 (2001): 16–21.

55 **on the Barker Hypothesis:** K. Calkins and S. U. Devaskar, "Fetal Origins of Adult Disease," *Current Problems in Pediatric Adolescent Health Care* 41 (2011): 158–76; D. J. Barker, P. D. Winter, C. Osmond, et al., "Fetal Nutrition and Cardiovascular Disease in Adult Life," *Lancet* 341 (1993): 938–41.

56 **hunger dominated all misery:** T. J. Roseboom, J. H. P. van der Meulen, A. C. Ravelli, et al., "Effects of Prenatal Exposure to the Dutch Famine on Adult Disease Later in Life: An Overview," *Molecular and Cell Endocrinology* 185 (2001): 93–98.

57 **on diabetes in Pima Indian women:** D. J. Pettitt and W. C. Knowler, "Diabetes and Obesity in the Pima Indians: A Cross Generational Vicious Cycle," *Obesity Weight Regulation* 7 (1988): 61–65; D. Dabelea,

R. L. Hanson, P. H. Bennett, et al., "Increased Prevalence of Type II Diabetes and Obesity in American Indian Children," *Diabetologia* 41 (1998): 904–10.

57 **slows growth in most organs:** J. Deardorff, "Your Health Partly Wired in the Womb," *Chicago Tribune,* November 13, 2011.

58 **two to six pounds of bacteria:** "NIH Human Microbiome Project Defines Normal Bacterial Makeup of the Body," *NIH News Release,* June 13, 2013, http://www.nih.gov/news/health/jun2012/nhgri-13.htm.

58 **on tooth decay:** M. Specter, "Germs Are Us," *New Yorker*¸ October 22, 2012, 32–39; R. Eckert, R. Sullivan, and W. Shi, "Targeted Antimicrobial Treatment to Re-establish a Healthy Microbial Flora for Long-Term Protection," *Advances in Dental Research* 24 (2012): 94–97.

59 **on the story of Alice:** A. Khoruts, J. Dicksved, J. K. Jansson, et al., "Changes in the composition of Human Fecal Microbiome after Bacteriotherapy for Recurrent *Clostridium difficile*-Associated Diarrhea," *Journal of Clinical Gastroenterology* 44 (2010): 354–60.

59 **tablet form for consumption:** I. Youngster, G. H. Russell, C. Pindar, et al., "Oral, Capsulized, Frozen Fecal Microbiota Transplantation for Relapsing *Clostridium difficile* Infection," *JAMA* 312 (2014): 1772–78; M. B. Smith, C. Kelly, and E. J. Alm, "How to Regulate Faecal Transplants," *Nature* 506 (2014): 290–91.

59 **evidence-based medical practice:** C. P. Kelly, "Fecal Microbiota Transplantation: An Old Therapy Comes to Age," *New England Journal of Medicine* 385 (2013): 474–75.

60 **fecal transplants or other means:** E. Yong, "There Is No 'Healthy' Microbiome," *New York Times,* November 2, 2014, 4.

60 **born by cesarean delivery:** J. A. Martin, B. E. Hamilton, M. J. K. Osterman, et al., "Births: Final Data for 2012," *National Vital Statistics Reports*, December 30, 2013, 2.

60 **on exposure to Staphylococcus via cesarean delivery:** M. G. Dominguez-Bello, E. K. Costello, M. Contreras, et al., "Delivery Mode Shapes the Acquisition and Structure of the Initial Microbiota Across Multiple Body Habitats in Newborns," *Proceedings of the National Academy of Science USA* 107 (2010): 11971–75.

60 **on the hygiene hypothesis:** G. A. W. Rook, "99th Dahlem Conference on Infection, Inflammation and Chronic Inflammatory Disorders: Darwinian Medicine and the 'Hygiene' or 'Old Friends' Hypothesis," *Clinical and Experimental Immunology* 160 (2010): 70–79.

61 **condition is 20 percent:** S. C. Fitzgibbons, Y. Ching, and D. Yu, "Mortality of Necrotizing Enterocolitis Expressed by Birth Weight Categories," *Journal of Pediatric Surgery* 44 (2009): 1072–75.

61 **on probiotics preventing NEC:** G. Deshpande, S. Rao, S. Patole, et al., "Updated Meta-analysis of Probiotics for Preventing Necrotizing Enterocolitis in Preterm Infants," *Pediatrics* 125 (2010): 921–30; K. Ganquli and W. A. Walker, "Probiotics in the Prevention of Necrotizing Enterocolitis," *Journal of Clinical Gastroenterology* 45 (2011): S133–38.

61 **the child is eighteen years old:** M. Sharland and the SACAR Pediatric Subgroup, "The Use of Antibacterials in Children: A Report of the Special

Advisory Committee on Antimicrobial Resistance (SACAR) Pediatric Subgroup," *Journal of Antimicrobial Chemotherapy* 60 suppl. (2007): i15–26.

61 **on H. pylori**: Y. Chen and M. J. Blaser, "Inverse Associations of Heliobacter pylori with Asthma and Allergy," *Archives of Internal Medicine* 167 (2007): 821–27; M. Blaser, "Antibiotic Overuse: Stop the Killing of Beneficial Bacteria," *Nature* 476 (2011): 393–94.

62 **steady, unassertive, and passive**: M. Usser, "Causes of Peptic Ulcers: A Selective Epidemiological Review," *Journal of Chronic Diseases* 20 (1967): 435–56.

62 **all those endoscopies go?**: "Barry J. Marshall," *NNDB: Tracking the Entire World*, http://www.nndb.com/people/899/000136491/.

63 **on earwax transplant**: Specter, "Germs Are Us."

64 **microbial wildlife managers**: Julie Segre, quoted in C. Zimmer, "Tending the Body's Microbial Garden," *New York Times*, June 18, 2012; see also W. P. Hanage, "Microbiome Science Needs a Healthy Dose of Skepticism," *Nature* 512 (2014): 247–48.

64 **dialogue with their surroundings**: Wilmut, Campbell, and Tudge, *The Second Creation*, 276.

Chapter 4: Pharmacogenomics

67 **measure of triumph**: Elie Wiesel, quoted in George J. Annas and Michael A. Grodin, *The Nazi Doctors and the Nuremberg Code* (New York: Oxford University Press, 1992), ix.

67 **on Henrietta Lacks**: Rebecca Skloot, *The Immortal Life of Henrietta Lacks* (New York: Random House, 2011).

68 **get some more blood**: Skloot, *The Immortal Life of Henrietta Lacks*, 190.

69 **testify to our common humanity**: Dorothy Roberts, *Fatal Invention* (New York: New Press, 2011), 258.

70 **delayed diagnosis and treatment**: B. R. Hunt, S. Whitman, and M. S. Hurlbert, "Increasing Black:White Disparities in Breast Cancer Mortality in the 50 Largest Cities in the United States," *Cancer Epidemiology* 38 (2014): 118–23.

70 **for all our citizens**: Bill Clinton, "Presidential Apology," White House Office of the Press Secretary, May, 16, 1997.

70 **in medical history**: James Jones, *Bad Blood: The Tuskegee Syphilis Experiment* (New York: Simon & Schuster, 1993), 91.

71 **cannot be undone**: H. Shaw, quoted in "Ceremony in Recognition of Survivors of the Study at Tuskegee, White House, East Room, May 16, 1997," *Tuskegee University YouTube Channel*, http://www.youtube.com/watch?v=l1A-YP24QwA.

71 **genetically conditioned drug reactions**: A. G. Motulsky, "Drug Reactions, Enzymes, and Biochemical Genetics," *Journal of American Medical Association* 165 (1957): 837.

72 **and colonized the world**: George J. Annas, *Worst Case Bioethics: Death, Disaster and Public Health* (New York: Oxford University Press, 2010), 254.

72 **believe we killed it:** E. Bonilla-Silva, quoted in Roberts, *Fatal Invention, (New York: The New Press, 2011)*, back cover.

73 **on BiDil:** Jonathan Kahn, *Race in a Bottle: The Story of BiDil* (New York: Columbia University Press, 2013), 2, 6.

73 **race in drug marketing:** Kahn, *Race in a Bottle*, 89, 109.

73 **Iraq or Afghanistan:** Nicolas Wade, *A Troublesome Inheritance: Genes, Race and Human History* (New York: Penguin, 2014), 127.

73 **American and Afghan societies:** H. A. Orr, "Stretch Genes," *New York Review of Books*, June 5, 2014, 61.

74 **on Serena Williams:** Serena Williams, quoted in M. Lauer, "Serena Williams Taking Recovery 'One Day at a Time,'" *Today Show*, March 9, 2011, http://www.nbcnews.com/id/21134540/vp/41986568#41986568.

75 **on Secretary Hillary Clinton:** H. Cooper and D. Grady, "Doctors Expect Clinton to Recover Fully from Blood Clot near Brain," *New York Times*, December 31, 2012.

76 **medicines Americans depend upon:** A. C. von Eschenbach, "FDA Approves Updated Warfarin (Coumadin) Prescribing Information," *New Genetic Information May Help Providers Improve Initial Dosing Estimates of the Anticoagulant for Individual Patients*, FDA News Release, August 16, 2007.

77 **first month of therapy:** S. E. Kimmel, B. French, and S. E. Kasner, "A Pharmacogenetic Versus a Clinical Algorithm for Warfarin Dosing," *New England Journal of Medicine* 369 (2013): 2283–93.

77 **on ultrarapid metabolizers:** M. Pirmohamed, "Pharmacogenetics: Past, Present, and Future," *Drug Discovery Today* 16 (2011): 853–54; G. Koren, J. Cairns, A. Gaedigk, et al., "Pharmacogenetics of Morphine Poisoning in a Breastfed Neonate of a Codeine-Prescribed Mother," *Lancet* 368 (2006): 704.

78 **on TPMT gene:** H. L. McLeod and C. Siva, "The Thiopurine S-methyltransferase Gene Locus—Implications for Clinical Pharmacogenomics," *Pharmacogenomics* 3 (2002): 89–98; N. K. DeBoer, A. A. van Bodegraven, P. de Graaf, et al., "Paradoxical Elevated Thiopurine S-methyltransferase Activity During Azathioprine Therapy: Potential Influence of Red Blood Cell Age," *Therapeutic Drug Monitoring* 30 (2008): 390–93.

79 **almanac-makers and the doctors:** Michel de Montaigne, *The Complete Essays of Montaigne*, trans. Donald Frame (Stanford, CA: Stanford University Press, 1958), 846–47; quoted in Stephen Greenblatt, *The Swerve* (New York: Norton, 2011), p. 244.

79 **G6PD deficiency in black prisoners:** R. S. Hockwald and C. B. Clayman, "Toxicity of Primaquine in Negroes," *JAMA* 149 (1952): 1568; M. D. Cappellini and G. Fiorelli, "Glucose 06-phosphate Dehydrogenase Deficiency," *Lancet* 371 (2008): 64–74.

80 **well-controlled studies:** "Companies Pitching Genetically Customized Nutritional Supplements Will Drop Misleading Disease Claims," *Federal Trade Commission*, January 7, 2014.

81 **opportunities for personalized medicine:** S. C. Sim and M. Ingelman-Sundberg, "Pharmacogenomic Biomarkers: New Tools in Current and Future Drug Therapy," *Trends in Pharmacological Sciences* 32 (2011): 72.

81 **deaths annually in the United States:** E. Hood, "Pharmacogenomics: The Promise of Personalized Medicine," *Environmental Health Perspectives* 111 (2003): A581; J. Lazarou, B. H. Pomeranz, and P. N. Corey, "Incidence of Adverse Drug Reactions in Hospitalized Patients: A Meta-Analysis of Prospective Studies," *JAMA* 279 (1998): 1200–1205.

82 **on Kalydeco:** B. Werth, "Cure for some could cost us all," *The Boston Globe,* March 23, 2014.

83 **contain pharmacogenomics information:** M. A. Hamburg and F. S. Collins, "The Path to Personalized Medicine," *New England Journal of Medicine* 363 (2010): 302.

83 **on clopidogrel poor metabolizers:** K. L. Hudson, "Genomics, Health Care, and Society," *New England Journal of Medicine* 365 (2011): 1037.

83 **on variants affecting clopidogrel response:** R. C. Green, H. L. Rehm, and I. S. Kohane, "Challenges in the Clinical Use of Genome Sequencing," in Geoffrey S. Gisburg and Huntington F. Ward, eds., *Genomics and Personalized Medicine* (New York: Academic Press, 2013), 112–13.

84 **advances in technology alone:** J. Sills, ed., "Personalize Medicine: Temper Expectations," in D. W. Nebert and G. Zhang, "Letters," *Science* 337 (2012): 910.

85 **well-designed randomized clinical trials:** S. E. Nissen, "Pharmacogenomics and Clopidogrel: Irrational Exuberance?" *JAMA* 306 (2011): 2727–28.

Chapter 5: Reprogenomics

87 **untested, unregulated treatments:** Debora L. Spar, *The Baby Business* (Boston: Harvard Business School, 2006), xiii.

87 **outside the reach of restrictive laws:** T. Audi and A. Chang, "Assembling the Global Baby," *Wall Street Journal,* December 10, 2010, C1.

88 **David Sigal and Brad Hoylman:** A. Hartocollis, "And Surrogacy Makes 3," *New York Times,* February 20, 2014, E1.

88 **on the case of Baby M:** George J. Annas, *Standard of Care* (New York: Oxford University Press, 1993), 61–70.

88 **traditional beginnings of a family:** S. K. Livio, "Christie Vetoes Bill That Would Have Eased Tough Rules for Gestational Surrogates," *Star-Ledger, Newark, NJ,* August 8, 2012.

89 **possibility of substitution disappears:** Spar, *The Baby Business,* 2, 4.

91 **Edwards Nobel Prize:** G. Vogel and M. Enserink, "Honor for Test Tube Bay Pioneer," *Science,* October 8, 2010, 158–59.

91 **on "Social Policy Considerations in Noncoital Reproduction":** S. Elias and G. J. Annas, "Social Policy Considerations in Noncoital Reproduction," *JAMA* 225 (1986): 62–68.

92 **by simply being:** "1978: Louise Brown," *People,* March 5, 1984 82.

94 **and criminal sanctions:** P. J. Williams, "Womb Wars," *Nation,* February 24, 2014, 10.

95 **California:** G. J. Annas, "Using Genes to Define Motherhood: The California Solution," *New England Journal of Medicine* 326 (1992): 417–20.

96 **on Canada's prohibition of sale of gametes and surrogacy**: G. J. Annas, "Assisted Reproduction: Canada's Supreme Court and the 'Global Baby,'" *New England Journal of Medicine* 365 (2011): 459–63, *and sources cited therein*. See also Michele Bratcher Goodwin, ed., *Baby Markets: Money and the New Politics of Creating Families* (New York: Cambridge University Press, 2010).

97 **It was uninheritable! But now?**: Carl Djerassi, *An Immaculate Misconception* (London: Imperial College Press, 2000).

98 **usefulness of egg freezing**: J. Johnston and M. Zoll, "Is Freezing Your Eggs Dangerous? A Primer," *New Republic*, November 1, 2014, www.newrepublic.com/article/120077/dangers-and-realities-egg-freezing.

99 **on Barbara Bailey**: S. S. Wang, "Genetic Tests for People Who'd Rather Not Know," *Wall Street Journal*, October 14, 2014, D3.

100 **annually with a mitochondrial disease**: A. M. Schaefer, R. W. Taylor, and D. M. Turnbull, "The Epidemiology of Mitochondrial Disorders—Past, Present and Future," *BBA: Bioenergetics* 1659 (2004): 115–20.

100 **on three-parent baby**: M. Tachibana, M. Sparman, H. Sritanaudomchai, et al., "Mitochondrial Gene Replacement and Embryonic Stem Cells," *Nature* 461 (2009): 367–72.

101 **gives the green light to tinker**: M. Le Page, Crossing the germ line—facing genetics' great taboo, *New Scientist*, February 6, 2015 (available at www.newscxientist.com/article/dn26927-crossing-the-germ-line-facing-genetics-greatest-taboo)

102 **"Can bioethics in the United States rise above politics?"**: G. J. Annas and S. Elias, "Politics, Morals and Embryos," *Nature* 431 (2004): 19–20.

103 **on ARTs statistics**: S. Sunderam, D. M. Kissin, S. B. Crawford, et al., "Assisted Reproductive Technology Surveillance—United States 2011," *Centers for Disease Control and Prevention Morbidity and Mortality Weekly Report* 63, November 21, 2014, 1–28.

104 **Cecil Jacobson**: A. Stone, "Va. Doctor Guilty in Fertility Fraud," *USA Today*, March 5, 1992, 1A; R. F. Howe, "Fertility Doctors' Ex-patients see Con Man, Saint," *Washington Post*, January 31, 1992, B1.

106 **for athletic ability**: Ricardo Asch, quoted in D. Sawyer, *Prime Time Live*, November 16, 1995.

106 **on Jamie Grifo**: A. Caplan, "Ethics and Octuplets: Society Is Responsible," *Philly.com*, February 6, 2009.

106 **on Michael Kamrava**: R. Lin, "Michael Kamrava's Medical License Revoked," *Los Angeles Times*, June 1, 2011, *http://documents.latimes.com/michael-kamrava-disciplinary-decision/*.

108 **a single healthy baby**: R. T. Scott and N. R. Treff, "Assessing the Reproductive Competence of Individual Embryos: A Proposal for the Validation of New '-Omics' Technologies," *Fertility and Sterility* 94 (2010): 791–94.

109 **has around 6.7 billion relatives**: A. Chakravarti, "Being Human: Kinship: Race Relations," *Nature* 457 (2009): 380–81.

Chapter 6: Genomic Messages from Fetuses

111 **fast in the past**: Jay Katz, *The Silent World of Doctor and Patient* (New York: Free Press, 1984), 202.

112 **a critically important option**: "ACMG Releases Statement on Access to Reproductive Options after Prenatal Diagnosis," *ACMG*, July 19, 2013, https://www.acmg.net/docs/Reproductive_Rights_News_Release.pdf.

112 **on North Dakota law: See generally** J. S. King, "Political and Fetal Diagnostics Collide," *Nature* 491 (2012): 33–34.

113 **boundaries of Roe v. Wade**: J. MacPherson, "North Dakota Abortion Ban Signed by Governor Jack Dalrymple," *Associated Press*, March 26, 2013.

113 **life or health**: *Roe v. Wade*, 410 U.S. 113 (1973) 165.

113 **terminate their pregnancy**: J. L. Natoli, D. L. Ackerman, S. McDermott, et al., "Parental Diagnosis of Down Syndrome: A Systematic Review of Termination Rates," *Prenatal Diagnosis* 32 (2012): 142–53.

114 **IOM clinical guidelines**: Lori Andrews, Jane Fullarton, Neil Holtzman, and Arno Motulsky, eds., *Assessing Genetic Risks: Implications for Health and Social Policy* (Washington, DC: National Academy Press, 1994), 7–8, 35.

115 **selection abortion**: Paul Billings, quoted in *Designing Genetic Information Policy: The Need for an Independent Policy Review of the Ethical, Legal, and Social Implications of the Human Genome Project, Sixteenth Report* (Washington, DC: U.S. Government Printing Office, 1992), 30.

116 **rates of prenatal screening and diagnosis**: "Births and Natality," *Centers for Disease Control and Prevention*, http://www.cdc.gov/nchs/fastats/births.htm.

118 **95 percent of all fetuses with Down syndrome**: F. D. Malone, J. A. Canick, and R. H. Ball, "First-Trimester or Second Trimester Screening, or Both, for Down's Syndrome," *New England Journal of Medicine* 353 (2010): 2001–11.

119 **on NIPS**: M. M. Gil, R. Akolekar, M. S. Quezada, et al., "Analysis of Cell-Free DNA in Maternal Blood in Screening for Aneuploidies: Meta-Analysis," *Fetal Diagnosis and Therapy* 35 (2014): 156–73.

123 **on the MAStudy**: R. J. Wapner, C. L. Martin, B. Levy, et al., "Chromosomal Microarray Versus Karyotyping for Prenatal Diagnosis," *New England Journal of Medicine* 367 (2012): 2175–84.

124 **on MAStudy interviews**: B. A. Bernhardt, D. Soucier, K. Hanson, et al., "Women's Experiences Receiving Abnormal Prenatal Chromosomal Microarray Testing Results," *Genetics in Medicine* 15 (2013): 139–45.

126 **on negative capability**: Katz, *The Silent World of Doctor and Patient*, 202–5.

127 **wood chipper**: Titus Brown, quoted in V. Marx, "The Genome Jigsaw," *Nature Reprints*, Tech. Features, November 2014, S39.

127 **for general obstetric patients**: "ACOG Committee Opinion No. 545, Noninvasive Prenatal Testing for Fetal Aneuploidy," *Obstetrics and Gynecology* 120 (2012): 1532–34.

128 **More recent data:** L. S. Chitty, "Use of Cell-Free DNA to Screen for Down's Syndrome," *New England Journal of Medicine,* DOI: 1-.1056/NEJMe1502441 (April 1, 2015); M. E. Norton et al., "Cell-Free DNA Analysis for Noninvasive Examination of Trisomy," *New England Journal of Medicine,* DOI: 1056/NEJMoa1407349 (April 1, 2015).

128 **reliable enough to make a diagnosis:** B. Daley, "Prenatal Tests Often Oversold, Imprecise," *Boston Globe,* December 14, 2014, A1.

129 **3,600 genes:** E. S. Lander, "Cutting the Gordian Helix: Regulating Genomic Testing in the Era of Precision Medicine," *New England Journal of Medicine* 372 (2015): 1185–86.

130 **warrant carrying to term:** E. Shuster, "Microarray Genetic Screening: A Prenatal Roadblock for Life?" *Lancet* 369 (2007): 526–29.

130 **gene police:** Georges Canguilhem, *The Normal and the Pathological* (New York: Zone Books, 1989), 280–81.

132 **on generic genetic consent:** S. Elias and G. J. Annas, "Generic Consent for Genetic Screening," *New England Journal of Medicine* 330 (1994): 1611–13.

136 **prenatal screening companies:** J. Mozersky, M. Mennuti, Cell-free fetal DNA testing: who is driving implementation? *Genetics in Medicine* 15 (2013) 433-4; and DNA Blood Test Gives Women a New Option for prenatal screening, NPR , January 28, 2015.

Chapter 7: Genomic Messages from Newborns and Children

139 **the sick and the well:** F. Scott Fitzgerald, *The Great Gatsby* (New York: Charles Scribner's Sons, 1925), 131.

141 **their consent obtained:** ACMG Board of Directors, "Points to Consider in the Clinical Application of Genomic Sequencing," *Genetics in Medicine* 14 (2012): 759–61; Committee on Genetics, "Molecular Genetic Testing in Pediatric Practice: A Subject Review," *Pediatrics* 106 (2000): 1494–97.

142 **withdrawal of life support:** C. J. Saunders, N. A. Miller, S. E. Soden, et al., "Rapid Whole-Genome Sequencing for Genetic Disease Diagnosis in Neonatal Intensive Care Units," *Science Translational Medicine* 4 (2012): 154ra135.

142 **single-gene genetic disorder:** Y. Yang, D. M. Muzny, and J. G. Reid, "Clinical Whole-Exome Sequencing for the Diagnosis of Mendelian Disorders," *New England Journal of Medicine* 369 (2013): 1502–11.

142 **5 percent of children:** Y. Yang, D. M. Muzny, F. Xia, et al., "Molecular Findings Among Patients Referred for Clinical Whole-Exome Sequencing," *JAMA* 312 (2014): 1870–79; H. Lee, J. L. Deignan, and N. Dorrani, "Clinical Exome Sequencing for Genetic Identification of Rare Mendelian Disorders," *JAMA* 312 (2014): 1880–87.

142 **applied more universally:** J. S. Berg, "Genome-Scale Sequencing in Clinical Care," *JAMA* 312 (2014): 1865–67.

142 **on Retta Beery:** Presidential Commission for the Study of Bioethical Issues, *Privacy and Progress in Whole Genome Sequencing* (October 2012), 14.

143 **on Nicholas Volker**: E. A. Worthey, A. N. Mayer, G. E. Syverson, et al., "Making a Definitive Diagnosis: Successful Clinical Application of Whole Exome Sequencing in a Child with Intractable Inflammatory Bowel Disease," *Genetics in Medicine* 13 (2011): 255–61.

144 **child's clinical outcome**: K. A. Johansen Taber, B. D. Dickinson, and M. Wilson, "The Promise and Challenges of Next-Generation Genome Sequencing for Clinical Care," *JAMA Internal Medicine* 174 (2014): 275–80.

144 **genetic advantage lies**: "What Is ACTN3 Sports Genet?" *Atlas Sports Genetics* (2009), http://www.atlasgene.com/.

144 **on ACTN3**: A. G. William and H. Wacherhage, "Genetic Testing in Athletes," *Medicine and Sports Science* 54 (2009): 174–86; N. Yang, D. G. MacArthur, and J. P. Gulbin, "*ACTN3* Genotype Is Associated with Human Elite Athletic Performance," *American Journal of Human Genetics* 73 (2003): 627–31.

145 **race the other kids**: David Epstein, *The Sports Gene* (New York: Current, 2013), 157.

146 **on ACMG recommendations**: Robert C. Green, Jonathan S. Berg, et al., *ACMG Recommendations for Reporting of Incidental Findings in Clinical Exome and Genome Sequencing* (American College of Medical Genetics and Genomics, 2013); ACMG Board of Directors, "ACMG Policy Statement: Updated Recommendations Regarding Analysis and Reporting of Secondary Findings in Clinical Genome-Scale Sequencing," *Genetics in Medicine,* November 13, 2014.

146 **respect for their rights**: S. M. Wolf, G. J. Annas, and S. Elias, "Patient Autonomy and Incidental Findings in Clinical Genomics," *Science* 340 (2013): 1049–50.

147 **nothing about their DNA**: H. Gilbert Welch, Lisa Schwartz, and Steven Woloshin, *Overdiagnosed* (Boston: Beacon Press, 2011), 134–35.

148 **complexity of clinical care**: W. G. Feero, "Clinical Application of Whole-Genome Sequencing: Proceed with Care," *JAMA* 311 (2014): 1017–19.

148 **Life expectancy, 33 years:** *Gattaca* (Columbia Pictures, 1997).

149 **genetic and endocrine disorders**: Centers for Disease Control and Prevention, "Ten Great Public Health Achievements—United States, 2001–2010," *Morbidity and Mortality Weekly Report (MMWR),* May 20, 2011, http://www.cdc.gov/mmwr/preview/mmwrthml/mm6019a5.htm.

150 **for very rare conditions**: Stefan Timmermans and Mara Buchbinder, *Saving Babies? The Consequences of Newborn Genetic Screening* (Chicago: University of Chicago Press, 2013), 233.

150 **for all babies everywhere**: D. B. Paul, "Appendix 5: The History of Newborn Phenylketonuria," in Neil Holtzman and M. S. Watson, eds., *Screening in the US, Promoting Safe and Effective Genetic Testing in the United States: Final Report of the Task Force on Genetic Testing* (September 1997), http://biotech.law.lsu.edu/research/fed/tfgt/appendix5.htm.

151 **lost to follow-up**: S. A. Berry, C. Brown, M. Grant, et al., "Newborn Screening 50 Years Later: Access Issues Faced by Adults with PKU," *Genetics in Medicine* 15 (2013): 591–99.

151 **on neonatal screening survey results:** R. Faden, A. J. Chwalow, N. A. Holtzman, et al., "A Survey to Evaluate Parental Consent as Public Policy

for Neonatal Screening," *American Journal of Public Health* 72 (1982): 1347–52.

151 **We think not:** R. R. Faden, N. A. Holtzman, and A. J. Chwalow, "Parental Rights, Child Welfare, and Public Health: The Case of PKU Screening," *American Journal of Public Health* 72 (1982): 1396–1400.

152 **genetic well-being:** G. J. Annas, "Mandatory PKU Screening: The Other Side of the Looking Glass," *American Journal of Public Health* 72 (1982): 1401–3.

153 **thirty-one primary conditions:** American College of Medical Genetics, "Newborn Screening: Towards a Uniform Screening Panel and System," *Genetics in Medicine* 8 (2006): S12–S252; "Recommended Uniform Screening Panel of the Secretary's Advisory Committee on Heritable Disorders in Newborns," *US Department of Health and Human Services*, http://www.hrsa.gov/advisorycommittees/mchbadvisory/heritabledisorders/recommendedpanel/index.html.

154 **on false-positive statistics:** A. Schulze, M. Lindner, D. Kohlmuller, et al., "Expanded Newborn Screening for Inborn Errors of Metabolism by Electrospray Ionization-Tandem Mass Spectrometry: Results, Outcome and Implications," *Pediatrics* 111 (2003): 1399–1406; C. Kwon and P. M. Farrell, "The Magnitude and Challenge of False-Positive Newborn Screening Test Results," *Archives of Pediatric and Adolescent Medicine* 154 (2000): 714–18. And see Appendix B.

154 **on impact of false positives on families:** J. L. Schmidt, K. Castellanos-Brown, S. Childress, et al., "The Impact of False-Positive Newborn Screening Results on Families: A Qualitative Study," *Genetics in Medicine* 14 (2012): 76–80.

156 **content is inconsistent:** A. R. Kemper, K. E. Fant, and S. J. Clark, "Informing Parents About Newborn Screening," *Public Health Nursing* 22 (2005): 332–38.

156 **and this makes sense:** Committee on Genetics, "Committee Opinion Number 481," *American College of Obstetricians and Gynecologists,* March 2011.

156 **approach to genetic screening:** President's Council on Bioethics, *The Changing Moral Focus of Newborn Screening* (December 2008), 51.

158 **more harm than good:** Karen Rothenberg and Lynn Bush, *The Drama of DNA: Narrative Genomics* (New York: Oxford University Press, 2014), 54.

158 **all of medicine is genomic medicine.** Robert C. Green, quoted in A. D. Marcus, "Scientists Will Study Genome Sequencing of Newborns," *Wall Street Journal,* December 30, 2014, D1.

159 **ethics and science right:** "Sequenced from the Start," *Nature* 501 (2013): 135; see also B. M. Knoppers, K. Senecal, P. Borry, and D. Avard, "Whole-Genome Sequencing in Newborn Screening Programs," *Science Translational Medicine* 6 (2014): 229.

Chapter 8: Cancer Genomics

161 **kingdom of the sick**: Susan Sontag, *Illness as Metaphor and AIDS and Its Metaphors* (New York: Macmillan, 1979), 3.

161 **difficult time of transition**: L. Gravitz, "Therapy: This Time It's Personal," *Nature* 509 (2014): S52–S54.

162 **but the body count**: Sontag, *Illness as Metaphor and AIDS and Its Metaphors,* 65.

162 **on cancer statistics**: "Cancer Facts & Figures 2013," *American Cancer Society*, http://www.cancer.org/research/cancerfactsstatistics/cancerfactsfigures2013/.

162 **majority of variation in cancers . . . due to "bad luck"**: C. Tomasetti, B. Vogelstein, "Variation in cancer risk among tissues can be explained by the number of stem cell divisions," *Science*, 347 (2015) 78-80. Also, J. Couzin-Frankel, "The bad luck of cancer," *Science*, 347 (2015) 12-13, and editorial, "Cancer: mixed messages, common purpose," *Lancet*, 385 (2015) 201.

164 **on the hypothetical Joan**: American Society of Clinical Oncology, *Accelerating Progress Against Cancer: ASCO's Blueprint for Transforming Clinical and Translational Cancer Research* (November 2011).

165 **on the hallmarks of cancer**: D. Hanahan and R. A. Weinberg, "The Hallmarks of Cancer," *Cell* 100 (2000): 57–70; D. Hanahan and R. A. Weinberg, "Hallmarks of Cancer: The Next Generation," *Cell* 144 (2011): 646–74.

168 **feeling like death**: Christopher Hitchens, *Mortality* (Toronto: McClelland & Stewart, 2012).

171 **fundamental understanding of cancer**: B. Vogelstein, N. Papadopoulos, and V. E. Velculescu, "Cancer Genome Landscapes," *Science* 339 (2013): 1546–58.

171 **Cancer Genome Atlas**: The Cancer Genome Atlas Research Network et al., "The Cancer Genome Atlas Pan-Cancer Analysis Project," *Nature Genetics* 45 (2013): 1113–20, and H. Ledford, End of cancer atlas prompts rethink, *Nature* 517 (2015) 128-9.

173 **FDA approves olaparib**: FDA Press Release, "FDA Approves Lynparza to Treat Advanced Ovarian Cancer," December 19, 2014, http://www.fda.gov/NewsEvents/Newsroom/PressAnnouncements/ucm427554.htm

173 **talented enough to do it**: K. Schulman, "President Obama on the Passing of Steve Jobs," *White House Blog*, October 5, 2011, http://www.whitehouse.gov/blog/2011/10/05/president-obama-passing-steve-jobs-he-changed-way-each-us-sees-world. .

174 **to die of it**: Walter Isaacson, *Steve Jobs* (New York: Simon & Schuster, 2011).

174 **5 percent of all new cancers**: D. Holmes, "The Cancer That Rises with the Sun," *Nature Outlook: Melanoma* 515 (2014): S110–11.

174 **well under one year**: D. B. Johnson and J. A. Sosman, "Update on the Targeted Therapy of Melanoma," *Current Treatment Options in Oncology* 14 (2013): 280–92.

174 **on BRAF gene**: H. Curtin, J. Fridlyand, T. Kageshita, et al., "Distinct Sets of Genetic Alterations in Melanoma," *New England Journal of Medicine*

353 (2005): 2135–47; C. Wellbrock, M. Karasarides, and R. Marais, "The RAF Proteins Take Center Stage," *Nature Reviews: Molecular Cell Biology* 5 (2004): 875–85.

175 **substantial percentage of patients**: K. S. M. Smalley and V. K. Sondak, "Melanoma—an Unlikely Poster Child for Personalized Cancer Therapy," *New England Journal of Medicine* 363 (2010): 876–78.

175 **squamous cell carcinomas**: F. Su, A. Viros, C. Milagre, et al., "RAS Mutations in Cutaneous Squamous-Cell Carcinomas in Patients Treated with BRAF Inhibitors," *New England Journal of Medicine* 366 (2012): 207–15.

176 **helps fight drug resistance**: K. T. Flaherty, J. R. Infante, A. Daud, et al., "Combined BRAF and MEK Inhibition in Melanoma with BRAF V600 Mutations," *New England Journal of Medicine* 367 (2012): 1694–1703.

176 **on Coley's toxin**: Editorial, "The Failure of the Erysipelas Toxins," *JAMA* 24 (1894): 919.

176 **therapy with immunotherapy**: Gravitz, "Therapy: This Time It's Personal."

177 **not be detected on scans**: O. Hamid, C. Robert, A. Daud, et al., "Safety and Tumor Responses with Lambrolizumab (anti-PD-1) in Melanoma," *New England Journal of Medicine* 369 (2013): 134–44.

177 **substantial proportion of patients**: J. D. Wolchok, H. Kluger, M. K. Callahan, et al., "Nivolumab Plus Ipilimumab in Advanced Melanoma," *New England Journal of Medicine* 369 (2013): 122–33.

177 **to the next level**: J. L. Riley, "Combination Checkpoint Blockade—Taking Melanoma Immunotherapy to the Next Level," *New England Journal of Medicine* 369 (2013): 187–89.

177 **on Steinman**: L. Gravitz, "A Fight for Life That United a Field," *Nature* 478 (2011): 163–64.

180 **approach to breast cancer prevention**: M. S. Ong and K. D. Mandl, "National Expenditure for False-Positive Mammograms and Breast Cancer Overdiagnoses Estimated at $4 Billion a Year," *Health Affairs* 34 (2015): 576–583.

181 **on BRCA statistics**: "ACOG: ACOG Practice Bulletin No. 103, April 2009 (Reaffirmed 2011). "Hereditary Breast and Ovarian Cancer Syndrome," *Gynecologic Oncology* 113 (2009): 6–11. Note: Other ranges and statistics outside of the ranges in this source have been published, but the ACOG numbers represent the most up-to-date research.

181 **designated 617delT**: C. M. Phelan, E. Kwan, E. Jack, et al., "A Low Frequency of Non-founder BRCA1 Mutations in Ashkenazi Jewish Breast-Ovarian Cancer Families," *Human Mutation* 20 (2002): 352–57.

181 **1–2 percent, respectively**: Y. C. Tai, S. Domcheck, G. Parmigiani, et al., "Breast Cancer Risk Among Male BRCA1 and BRCA2 Mutation Carriers," *Journal of the National Cancer Institute* 99 (2007): 1811–14.

181 **risk of prostate cancer**: S. M. Edwards, Z. Kote-Jarai, J. Meitz, et al., "Two Percent of Men with Early-Onset Prostate Cancer Harbor Germline Mutations in the BRCA2 Gene," *American Journal of Human Genetics* 72 (2003): 1–12.

182 **National Cancer Institute's recommendations:** "BRCA1 and BRCA2: Cancer Risk and Genetic Testing," *National Cancer Institute,* http://www.cancer.gov/cancertopics/factsheet/Risk/BRCA

183 **on women of Ashkenazi background:** R. Caryn Rabin, "Study of Jewish Women Shows Link to Cancer Without Family History," *New York Times,* September 2, 2014, A15; and B. Daly, O. Olopade, Race, Ethnicity, and the Diagnosis of Breast Cancer, *JAMA* 313 (2015) 141-2.

183 **her own experiences:** Jennifer Couzin-Frankel, "Unknown Significance," *Science* 346 (2014): 1167–70.

185 **on VUS:** T. S. Frank, A. M. Deffenbaugh, J. E. Reid, et al., "Clinical Characteristics of Individuals with Germline Mutations in BRCA1 and BRCA2: Analysis of 10,000 Individuals," *Journal of Clinical Oncology* 20 (2002): 1480–90; J. N. Weitzel, V. Lagos, K. R. Blazer, et al., "Prevalence of BRCA Mutations and Founder Effect in High-Risk Hispanic Families," *Cancer Epidemiology, Biomarkers & Prevention* 14 (2005): 1666–71; R. Nada, L. P. Shumm, S. Cummings, et al., "Genetic Testing in an Ethnically Diverse Cohort of High-Risk Women: A Comparative Analysis of BRCA1 and BRCA2 Mutations in American Families of European and African Ancestry," *JAMA* 294 (2005): 1925–33.

185 **95 percent or higher:** Meijers-Heijboer, H., van Geel, B., van Putten, W.L., et al., "Breast Cancer after Prophylactic Bilateral Mastectomy in Women with BRCA1 or BRCA2 Mutations," *New England Journal of Medicine* 345 (2001): 159–64; L. C. Hartmann, T. A. Sellers, D. J. Schaid, et al., "Efficacy of Bilateral Prophylactic Mastectomy in BRCA1 and BRCA2 Gene Mutation Carriers," *Journal of the National Cancer Institute* 93 (2001): 1633–37.

185 **40–70 percent:** N. D. Kauff, J. M. Satagopan, M. E. Robson, et al., "Risk Reducing Salpingo-Oophorectomy in Women with a BRCA1 or BRCA2 Mutation," *New England Journal of Medicine* 346 (2002): 1609–15; N. D. Kauff, S. M. Domcheck, T.M. Friebel, et al., "Risk-Reducing Salpingo-Oophorectomy for the Prevention of BRCA1 and BRCA2-Associated Breast and Gynecologic Cancer: A Multicenter Prospective Study," *Journal of Clinical Oncology* 26 (2008): 1331–37; J. L. Kramer, I. A. Velazquez, B. E. Chen, et al., "Prophylactic Oophorectomy Reduces Breast Cancer Penetrance During Prospective, Long-Term Follow-Up of BRCA1 Mutation Carriers," *Journal of Clinical Oncology* 23 (2005): 8629–35.

185 **90 percent or higher:** T. R. Rebbeck, N. D. Kauff, and S. M. Domcheck, "Meta-Analysis of Risk Reduction Estimates Associated with Risk-Reducing Salpingo-Oophorectomy in BRCA1 or BRCA2 Mutation Carriers," *Journal of the National Cancer Institute* 101 (2009): 80–87.

186 **on the story of Emily:** N. Tung, "Management of Women with *BRCA* Mutations: A 41-Year Old Woman with a *BRCA* Mutation and Recent History of Breast Cancer," *JAMA* 305 (2011): 2211–20.

187 **As Frances Visco:** F. Visco, "Can We Beat Cancer at Its Own Game?" *New York Times,* April 4, 2015, A16. See also M. A. Sekeres, "Trying to Fool Cancer," *New York Times,* March 29, 2015, SR7.

188 **and therapeutic intervention:** Siddhartha Mukherjee, *The Emperor of All Maladies* (New York: Scribner, 2010), 458. See also *Cancer: The*

Emperor of all Maladies, PBS video, 2015, http://video.pbs.org/program/story-cancer-emperor-all-maladies/.

Chapter 9: Genomic Privacy and DNA Data Banks

191 **specific patterns and relationships:** Hallam Stevens, *Life Out of Sequence* (Chicago: University of Chicago Press, 2013), 204.

192 **special issue on "The End of Privacy":** *Science*, 347 (2015) 455.

192 **patient privacy is built into our efforts from Day I:** quoted in R. Pear, U.S. to Collect Genetic Data to Hone Care, *New York Times*, January 31, 2015, A11.

194 **be preserved:** George J. Annas and Sherman Elias, *Gene Mapping: Using Law and Ethics as Guides* (New York: Oxford University Press, 1992), 272.

196 **on Alonzo King:** Maryland v. King, 569 U.S. 1962 (2013).

197 **The Blooding:** Joseph Wambaugh, *The Blooding* (New York: Bantam Books, 1989).

197 **crime and terrorism:** G. J. Annas, "Protecting Privacy and the Public: Limits on Police Use of Bioidentifiers in Europe," *New England Journal of Medicine* 361 (2009): 196–201, and sources cited therein.

198 **right to be forgotten:** Google Inc. v Agencia Española de Protección de Datos, Mario Costeja Gonzalez, ECLI:EU:C:2014: 317

198 **Privacy is Theft:** Dave Eggers, *The Circle* (New York: Random House, 2013).

198 **Riley v. California:** Riley v. California, 573 U.S. (2014).

199 **these people's authority:** Alexandr Solzhenitsyn, *The Cancer Ward* (London: Random House, 2003), 208–9.

199 **ways to barcode them:** Editorial, "Watching Big Brother," *Nature* 456 (2008): 675–76.

200 **bodies and disease:** Stevens, *Life Out of Sequence*, 8, 204.

200 **via the Iceland legislature:** J. R. Gulcher and K. Stefansson, "The Icelandic Healthcare Data Base and Informed Consent," *New England Journal of Medicine* 342 (2000): 1827–30; G. J. Annas, "Rules for Research on Human Genetic Variation—Lessons from Iceland," *New England Journal of Medicine* 342 (2000): 1830–33.

200 ***Nature Genetics:*** D. F. Gudbjartsson et al., "Large-Scale Whole-Genome Sequencing of the Icelandic Population," *Nature Genetics* doi:10.1038/ng.3247, March 25, 2015.

200 **At the push of a button:** C. Zimmer, "Snapshot of Icelanders' DNA Reveals Gene Mutations Tied to Disease," *New York Times*, March 26, 2015, A6.

201 **traced to specific individuals:** Guomundsdottir v. Iceland, Icelandic Supreme Court, No. 151/2003, discussed in G. J. Annas, "Family Privacy and Death: Antigone, War, and Medical Research," *New England Journal of Medicine* (2005); 501–5.

202 **It just doesn't matter:** Quoted in M. Specter, "The Gene Factory," *New Yorker*, January 6, 2014, 40.

202 **even after death:** "An Afternoon at UK Biobank," *Lancet* 373 (2009): 1146.

202 **to share their genome**: George Church, quoted in "Welcome to My Genome," *Economist*, September 6, 2014, 21.

203 **Genetic Privacy Act**: G. J. Annas, L. H. Glantz, and P. A. Roche, "The Genetic Privacy Act and Commentary," February 28, 1995 (Sherman was the scientific adviser during the drafting process).

204 **differing privacy protections**: Editorial, "Ethical Overkill," *Nature* 516 (2014): 143–44.

204 **respected and cherished**: Editorial, "Careless Data," *Nature* 507 (2014): 7.

205 **plague Big Data**: M. J. Khoury and J. P. A. Ioannidis, "Big Data Meets Public Health," *Science* 346 (2014): 1054.

205 **several days**: D. Butler, "When Google Got Flu Wrong," *Nature* 494 (2013): 155–56.

205 **factors than the Google model**: NAHAM Government Relations, "Google Flu Trends Will Not Replace the CDC Flu Model," *National Association of Healthcare Access Management*, http://nahamnews.blogspot.com/2013/01/google-flu-trends-will-not-replace-cdc.html.

205 **even the best data tools**: A. O'Leary, "In New Tools to Combat Epidemics, the Key Is Context," *New York Times*, June 20, 2013, F2.

206 **on John Moore's case**: George J. Annas, *Standard of Care* (New York: Oxford University Press, 1993), 167–77; *Moore v. Regents of the University of California*, 793 P.2d 479, 271 Cal. Rptr. 146 (1990).

207 **cripple medical research**: *Greenberg v. Miami Children's Research Institute*, 264 F. Supp. 2d 1064 (S.D. Fla. 2003).

207 **tissue to the researchers**: Washington University v. Catalona, 490 F.3d 667 (8th Cir. 2007). On new tissue rules, see generally L. Glantz, P. Roche, and G. J. Annas, "Rules for Donations to Tissue Banks: What Next?" *New England Journal of Medicine* 358 (2008): 298–303.

209 **workable consent process**: Presidential Commission for the Study of Bioethical Issues, *Privacy and Progress in Whole Genome Sequencing* (October 2012), 5, 79, 91.

210 **participates in research**: Jeri Lacks-Whye, quoted in E. Callaway, "Deal Done over HeLa Cell Line," *Nature* 500 (2013): 132–33.

210 **capacity for degrading oil**: *Diamond v. Chakrabarty*, 447 US 303 (1980).

211 **Too bad**: Michael Crichton, "Patenting Life," *New York Times*, February 13, 2007, A23.

211 **on patentable breast cancer genes**: *Association for Molecular Pathology, et al. v Myriad Genetics, Inc.*, 132 S. Ct. 1794 (2012).

213 **sharing their data**: Presidential Commission for the Study of Bioethical Issues, *Privacy and Progress in Whole Genome Sequencing*, 3.

214 **can be identifiable**: F. H. Cate, "The Big Data Debate," *Science* 346 (2014): 818.

Chapter 10: Genomics Future

217 **of course, science fiction**: Craig Venter, *Life at the Speed of Light: From the Double Helix to the Dawn of the Digital Age* (New York: Viking, 2013), 160.

218 **the century of biology:** F. Dyson, "Our Biotech Future," *New York Review of Books,* July 19, 2007, 54.

219 **perfectly adjusted to their environment:** Margaret Atwood, *Oryx and Crake* (New York: Anchor Books, 2004), 305.

219 **end by killing themselves:** Margaret Atwood, *MaddAddam* (New York: Doubleday, 2009), 291.

220 **obtained from fresh DNA:** Venter, *Life at the Speed of Light,* 87.

221 **Gone but not Forgotten:** K. Higgins, "Damien Hirst Unveils Gilded Woolly Mammoth Skeleton with 'Gone but Not Forgotten,'" *Greenlabel.com,* http://green-label.com/art/damien-hirst-gilded-woolly-mammoth-skeleton-gone-but-not-forgotten/.

222 **become a political force** and *their facial morphology* and *laws can change:* "Interview with George Church: Can Neanderthals Be Brought Back from the Dead?" *Der Spiegel,* January 14, 2013, http://www.spiegel.de/international/zeitgeist/george-church-explains-how-dna-will-be-construction-material-of-the-future-a-877634.html.

222 **and is widely used:** George Church and Ed Regis, *Regenesis: How Synthetic Biology Will Reinvent Nature and Ourselves* (New York: Basic Books, 2012), 148.

222 **cloned cave baby:** A. Hall and F. Macrae, "Wanted: 'Adventurous Woman' to Give Birth to Neanderthal Man—Harvard Professor Seeks Mother for Cloned Cave Baby," *Daily Mail,* January 20, 2013, http://www.dailymail.co.uk/news/article-2265402/Adventurous-human-woman-wanted-birth-Neanderthal-man-Harvard-professor.html.

222 **has never been so apt:** J. Naish, "Neanderthal Man, DNA Experiments and the Shadow of Dr. Frankenstein," *Daily Mail,* January 21, 2013, http://www.dailymail.co.uk/sciencetech/article-2266108/Neanderthal-man-DNA-experiments-shadow-Dr-Frankenstein.html.

224 **the history of science:** Michael Crichton, *Jurassic Park* (New York: Knopf, 1990), x.

224 **Science writer Barry Werth:** Barry Werth, *The Billion Dollar Molecule* (New York: Simon and Schuster) 1994, 189.

225 **ethical arbitrage:** R. M. Isasi and G. J. Annas, "Arbitrage, Bioethics, and Cloning: The ABCs of Gestating a United Nations Cloning Convention", *Case Western Reserve Journal of International Law* 35 (2003): 397–414.

227 **ending life on earth:** Kurt Vonnegut, *Cat's Cradle* (New York: Random House, 2010), 39.

227 **genetic genocide:** George J. Annas, "Genism, Racism and the Prospect of Genetic Genocide," in Jérôme Bindé, ed., *The Future of Values: 21st Century Talks* (New York: Berghahn, 2004), 284.

228 **are like as a species:** Francis Fukuyama, *Our Posthuman Future* (New York: Macmillan, 2002), 128.

228 **based on becoming immortal:** J. Lanier, *Who Owns the Future?* (New York: Simon & Schuster, 2013), 326.

228 **megawizard in futurist circles:** J. Maslin, "Fightin Words Against Big Data," *New York Times,* May 6, 2013, C1.

229 **on "Can Google Solve Death?":** H. McCracken and L. Grossman, "Google vs. Death," *Time,* September 30, 2013, 24–30.

229 **This baby could live to be 142:** *Time* cover, Special Health Double Issue, February 23-March 2, 2015.
230 **on Baseline Study:** A. Barr, "Google's New Moonshot Project: The Human Body," *Wall Street Journal*, July 27, 2014.
231 **Kepler space observatory:** Venter, *Life at the Speed of Light*, 187.
232 **long line of human advances:** Dan Brown, *Inferno* (New York: Knopf, 2013).
232 **the Singularity:** Ray Kurzweil, *The Singularity is Near* (New York, Viking) 2007, 7.
232 **will trump technology:** J. Enriquez and S. Gullans, *Homo Evolutis: Please Meet the Next Human Species* (New York: Ted Conferences, 2011).
233 **risky and nontherapeutic:** E. Lanphier et al., "Don't edit the human germ line," *Nature* 519 (2015): 410.
233 **We are humans:** Quoted in D. Cyrandski, "Embryo editing divides scientists," *Nature* 519 (2015): 272.
233 **need for open discussion:** D. Baltimore et al., "A prudent path forward for genomic engineering and germline gene modification," *Science* 348 (2015): 36–38.
234 **needs to improve:** T. Frieden, quoted in "CDC Press Conference on Laboratory Quality and Safety after Recent Lab Incidents," *Center for Disease Control*, July 11, 2014, ttp://www.cdc.gov/media/releases/2014/t0711-lab-safety.html.
234 **no such thing as perfect systems :** Thomas Inglesby quoted in J. Cohen, "Alarm over Biosafety Blunders," *Science* 345 (2014): 247–48.
234 **as an acceptable threshold:** Martin Rees, *Our Final Hour* (New York: Basic Books, 2003), 127.
235 **on McConnell:** T. McConnell, "Genetic Enhancement, Human Nature, and Rights," *Journal of Medical Philosophy* 35 (2010): 415–28.
235 **on Eric Juengst:** E.T. Juengst, "Group Identity and Human Diversity: Keeping Biology Straight from Culture," *Human Genetics* 63 (1998): 673-677.
236 **size of the primal tribe:** Atwood, *Oryx and Crake*, 343.
236 **not possible in theory:** Atwood, *MaddAddam*, 393.
236 **science slightly over the edge:** Ed Regis, *Great Mambo Chicken and the Transhuman Condition* (New York: Addison-Wesley, 1990).
237 **genome of the Universe:** Church and Regis, *Regenesis*, 25–26, 237–39.
237 **and omnipresent:** Frank Tipler, *The Physics of Christianity* (New York: Doubleday, 2007).

Appendix B: Limitations of Screening Tests

249 **benefit from screening:** S.D. Saini, F. van Hees, S. Vijan, "Smarter screening for cancer," *JAMA* 312 (2014): 2211–12.
251 **is about 0.01 percent:** *Omni Whole-Genome DNA Analysis BeadChips*, San Diego: Illumina Inc. (2013).
251 **10 times to 30 times:** *Estimating Sequencing Coverage*, San Diego: Illumina Inc. (2011).

251 **of their interesting SNPs:** L. Pachter, "Multiple testing an issue for 23andMe," *Bits of DNA*, November 30, 2013.

251 **coverage depth (less than 5 times):** D. Shigemizu, A. Fujimoto, S. Akiyama, et al., "A practical method to detect SNVs and indels from whole genome and exome sequencing data," *Nature Scientific Reports* 3 (2013): 2161.

251 **from 450 to 11:** H., Ledford, "Lists of Cancer Mutations Awash with False Positives," *Nature News*, June 17, 2013; M.S. Lawrence, P. Stojanov, P. Polak, et al., "Mutational Heterogeneity in Cancer and the Search for New Cancer-associated Genes," *Nature* 499 (2013): 214–218.

251 **six were consistently replicated:** J.N. Hirschhorn, K. Lohmueller, E. Byrne, et al., "A Comprehensive Review of Genetic Association Studies," *Genetics in Medicine* 4 (2002): 45–61.

251 **cell free tumor DNA:** C. Curtin, "At ACMG, Presenters Discuss False-positive, False-negative NIPT Results," *GenomeWeb News*, March 25, 2013.

252 **direct applicability to genomics:** Nate Silver, *The Signal and the Noise* (New York: Penguin, 2012).

252 **all at high risk of disease:** H. Gilbert Welch, Lisa Schwartz, and Steven Woloshin, *Overdiagnosed* (Boston: Beacon Press, 2011), 135.

Glossary

ACMG: American College of Medical Genetics and Genomics.

Amniocentesis: A prenatal test in which a small amount of amniotic fluid is removed from the amniotic sac surrounding the fetus and cells within the fluid are tested for genomic abnormalities.

Alleles: Varieties or alternatives of a single gene. An individual inherits two alleles from each parent and which alleles a person inherits may affect the physical traits they express.

ART: Assisted Reproduction Technology.

Carrier: An individual who is heterozygous for a given trait that only affects individuals who are homozygous. Thus, although carriers do not display any physical characteristics of the trait, they can pass on the trait to their offspring.

Chromosomes: The packaged structure of DNA within the nucleus of a cell. Humans have 46 chromosomes.

Clone: A genetic duplicate. This may refer to an organism or a DNA fragment.

CMA (chromosomal microarray): A high-volume, automated analysis used to identify duplicated or deleted copies of genetic data (CNVs). The analysis compares a patient's DNA to a reference genome using small DNA probes.

CNVs (copy number variants): Gains, losses, and inversions of stretches of DNA larger than 1,000 base pairs. Most CNVs are normal and contribute to human variation.

CVS (chorionic villus sampling): A prenatal test in which cells from the placenta of a fetus are removed and tested for genomic abnormalities.

Deletion: The loss of genetic material.

DNA (deoxyribose nucleic acid): A large organic molecule made up of four bases (*A*denine, *G*uanine, *C*ytosine, and *T*hymine) that encode all the basic information necessary for life. DNA is double stranded, resulting in base pairs between strands, and has a helical structure.

Dominant: An allele that masks the presence of a recessive allele and is expressed when the individual has one (heterozygous dominant) or two (homozygous dominant) copies of a dominant allele.

EHR: Electronic Health Records (sometimes termed electronic medical records).

ENCODE: *(n)* Encyclopedia Of DNA Elements, a project to determine all the functional elements in the human genome sequence.

Encode: *(v)* To convert from one system of communication (DNA base pairs) to another system of communication (amino acids in a protein chain).

Epigenetics: "On top of genetics", the study of chemical and structural changes to the genome.

Epigenome: All chemical and structural changes to the genome.

Eugenics: The belief and practice of improving the genetic quality of the human species, generally through the selective reproduction of individuals with favorable traits and the (generally involuntary) diminished reproduction of individuals with less favorable traits.

EWAS (epigenome-wide association studies): Studies in which the epigenomes of ill and healthy individuals are compared to identify genetic modifications associated with disease. Current research specifically focuses on the addition of a carbon-hydrogen molecule (methyl) to DNA.

Exome: The part of the genome comprised of protein coding genes. This constitutes less than 2 percent of the total genome.

False negative: When a test result is negative, but the patient does have the condition being tested for.

False positive: When a test result is positive, but the patient does not actually have the condition being tested for.

FDA: U.S. Food and Drug Administration.

Gene: A segment of DNA base pairs that contain the information necessary to encode a protein (or polypeptide chain).

Genetic counseling: A practice in which patients and family members receive counseling regarding the nature and consequences of a genetic disorder.

Genetics: The study of the actions and structure of genes, as well as the patterns of inheritance through which genes are passed on.

Genism: The belief that fundamental human characteristics are determined by genes.

Genome: All genetic components of a single individual or species.

Genotype: The genetic makeup of an individual.

GWAS (genome-wide association studies): Studies in which the genomes of ill and healthy people are compared to identify genetic variations associated with disease.

Heterozygous: When an individual inherits two different alleles of a given gene from their biological mother and father.

Homozygous: When an individual inherits the same alleles of a given gene from their biological mother and father.

Human Genome Project: An international scientific project to determine the chemical makeup of human DNA, including all human genes and their functions.

ICSI: Intra-cytoplasmic sperm injection.

IVF: *In vitro* fertilization.

Mendelian inheritance: A pattern of inheritance characteristic of organisms that reproduce sexually. Of greatest importance is Mendel's Law of Dominance, which states that some alleles are dominant while others are recessive.

Microbiome: The ecological community of microorganisms that share our body space.

Mutation: A change in genetic material such that a new variation is produced. Mutations can occur through insertions of new base pairs, deletions of base pairs, inversions of stretches of DNA, etc.

NIH: U.S. National Institutes of Health.

Oncogene: Genes that increase the likelihood that a cell will become cancerous due to uncontrolled growth.

Pedigree: A diagram of heredity that traces a trait or disease of interest through many generations.

Personalized medicine: A model that advocates medical care customized to an individual patient's unique cellular and genetic makeup.

Phenotype: The physical makeup of an individual as determined by their genotype.

Protein: A large organic molecule encoded by DNA that interacts with enzymes, performs important cellular metabolic functions, and supports cells structurally.

Positive predictive value: The probability that a patient with a positive test result does in fact have the disease being tested for. This is calculated by dividing the number of true positive test results by the total number of positive tests results.

Racism: The belief that fundamental human characteristics are determined by race.

Recessive: An allele that is masked by the presence of a dominant allele and only manifests in individuals with two recessive alleles (homozygous recessive).

Sensitivity: The probability of a positive test among patients with the disease.

Singularity: The hypothetical moment when artificial intelligence is so advanced that humanity will undergo a dramatic and irrevocable change.

SNPs (single nucleotide polymorphisms): A single base pair variation in a DNA sequence. For example, a fragment of DNA from one individual may read AACG**T**CG, while a fragment of DNA from another individual may read AACG**C**CG.

Specificity: The probability of a negative test among patients without the disease.

Tumor suppressor gene: Genes that normally restrict the development of tumorous cells. If the tumor suppressor genes of a cell stop functioning, the cell is more likely to become cancerous.

Variants of uncertain significance (VUS): A variant identified in an individual's genome that cannot be classified as benign or pathogenic due to a lack of information about that particular variant.

WES: Whole-exome sequencing.

WGS: Whole-genome sequencing.

Index

Page numbers in *italics* refer to illustrations.